Oil Prices and the Future of OPEC

The policy of the United States and, by extension, that of many oil importing countries, toward OPEC countries is in large part a function of an estimate of the factors that condition oil decisions in exporting countries. In this title, originally published in 1978, Ted Moran examines how immune OPEC can expect to be to the struggles over market shares that traditionally have beset attempts to organize natural resource cartels. Moran's research leads him to argue that skyrocketing commitments to growth and social betterment leave little slack in national budgets and thus preclude output reductions for any extended period, or at least act as a substantial deterrent, unless such reductions come in support of an effort to raise real oil prices substantially. For any student interested in international policy, economic development, or environmental studies, this title offers fascinating insights into the oil industry.

Oil Prices and the Future of OPEC

The Political Economy of Tension and Stability in the
Organization of Petroleum Exporting Countries

Theodore H. Moran

RFF PRESS
RESOURCES FOR THE FUTURE

First published in 1978
by Resources for the Future, Inc.

This edition first published in 2016 by Routledge
2 Park Square, Milton Park, Abingdon, Oxon, OX14 4RN
and by Routledge
711 Third Avenue, New York, NY 10017

Routledge is an imprint of the Taylor & Francis Group, an informa business

© 1978, Resources for the Future, Inc.

Publisher's Note
The publisher has gone to great lengths to ensure the quality of this reprint but points out that some
imperfections in the original copies may be apparent.

Disclaimer
The publisher has made every effort to trace copyright holders and welcomes correspondence from
those they have been unable to contact.

A Library of Congress record exists under LC control number: 78002983

ISBN 13: 978-1-138-95420-5 (hbk)
ISBN 13: 978-1-315-66705-8 (ebk)
ISBN 13: 978-1-138-95421-2 (pbk)

Oil prices and the future of OPEC

The political economy of tension and stability
in the Organization of Petroleum Exporting Countries

THEODORE H. MORAN

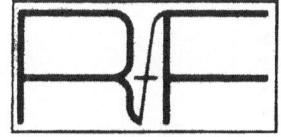

RESEARCH PAPER R-8

RESOURCES FOR THE FUTURE / WASHINGTON, D.C.

Library of Congress Catalog Card Number 78-2983
ISBN 0-8018-2117-7

Manufactured in the United States of America

Published March 1978.

The policy of the United States and, by extension, that of many other oil importing countries, toward the OPEC countries is in large part a function of an estimate of the factors that condition oil output decisions in the exporting countries.

The specific thrust of Ted Moran's inquiry may be formulated as "how immune can OPEC expect to be to the struggles over market shares that traditionally have beset attempts to organize natural resource cartels?" There exists a wide range of judgment on this. One widespread opinion is that OPEC will always have an easy job of allocating demand among its members because the key producers in the Persian Gulf can cut back production drastically, if need be, and act as residual suppliers because they do not need the revenues. Others argue that skyrocketing commitments to growth and social betterment leave little slack in the national budgets and thus preclude output reductions for any extended period, or at least act as a substantial deterrent, unless such reductions come in support of an effort to raise real oil prices substantially.

Moran's research has led him to favor the latter conclusion. He arrives at it through painstaking evaluation of projected development, leading him to reassess the likely costs at a much higher level than put down in official plans. This is tedious work. Moreover, it is never fully subject to convincing proof. We are in the author's debt for having taken on the task.

As always, RFF only guarantees the craftsmanship; it does not "underwrite" the findings. The reader will have to evaluate those in the light of other experts' opinion and of evolving conditions in the Middle East. As for the author, we like to believe that the work done by him under the RFF grant has served him well to prepare him for his current responsibility.

This research was funded by a grant from Resources for the Future. Dr. Moran conducted the study while he was a visiting associate professor at the Johns Hopkins University School of Advanced International Studies. He was aided by the research assistance of Katherine Valyi and Darcie Bundy. Dr. Moran is currently a member of the Policy Planning Staff, U.S. Department of State.

February 1978

Hans H. Landsberg
Co-Director,
Center for Energy Policy Research

Table of Contents

I. INTRODUCTION

The Organization of Petroleum Exporting Countries (OPEC) faces the
organizational problem common to all oligopolies: the economic benefits
to the group as a whole will be maximized if the members coordinate
production decisions as if they were a monopolistic supplier, but the
economic benefits to individual members will be maximized if they can
expand their output by offering price discounts without being disciplined
by the cartel and without inducing the cartel to fall apart. When, as
in the case of OPEC, the demand for the producer group's output is relatively
inelastic while the demand for any individual producer's output is highly
elastic, and the marginal production costs are small compared to the
group's asking price, the rewards for cheating are great. As a consequence,
to maintain the oligopoly, each member must exercise self-discipline in
the common good and be assured that his fellow members will do likewise.
This requires some (explicit or implicit) agreement on the distribution of
market shares and some method of monitoring and enforcing that distribution.

For the past few years, OPEC has had a special advantage in dealing
with the question of market shares and with the problem of cheating: the
member states with the greatest ability to expand output have not "needed"
the revenues that additional production could have generated; the member
governments that have "needed" the revenues the most have been able to
produce at near capacity. Those production cutbacks that have been
necessary to balance supply and demand at the prices dictated by OPEC have,
in general, been able to be shunted off onto low-population, low-mobili-
zation countries for whom the marginal utility of the revenues foregone
has been, by all objective measurements, low. Mild cases of cheating on

the part of some of the high-population, high-mobilization states have been tolerated by the other OPEC members with only minor acrimony. The tension within OPEC has remained at a low level. There has been no need to decide upon, or to monitor, or to enforce an explicit distribution of market shares.

How far will this blessing of "easy adjustment" persist into the future? Will OPEC ever have to face the problem that has bedeviled other oligopolies—the prospect that members with both a strong incentive to cheat and a great capability to do so will have to hold extensive amounts of producing facilities permanently idle? How might OPEC react to rising tension about the distribution of market shares among its members?

These are questions that have received scant attention.[1] Studies that have disaggregated the simulation of OPEC behavior below the level of how a perfect monopolist might act in the abstract typically hypothesize two sub-groups within OPEC: a "competitive fringe" of high population countries that continuously produces close to capacity, and a "core of balancers" that adjusts production to meet the remaining demand. A common starting point, in fact, seems to be that Saudi Arabia alone can act as residual supplier, cutting back whatever exports are necessary to balance supply and demand at the price chosen by OPEC.

The assumption embodied in the conventional approach to modeling OPEC behavior—namely, that the question of the internal distribution of the OPEC

---

[1]For modeling of OPEC behavior in relation to the international energy market, see C. Blitzer, A. Meeraus, and A. Stoutjesdijk, "A Dynamic Model of OPEC Trade and Production," _Journal of Development Economics_, Fall 1975; D. Fischer, D. Gately, and J. F. Kyle, "The Prospects for OPEC: A Critical Survey of Models of the World Oil Market," _Journal of Development Economics_ vol. 2, 1975; E. Hnyilicza and R. S. Pindyck, "Pricing Policies for a Two-Part Exhuastible Resource Cartel: The Case of OPEC," _European Economic Review_, to appear; B. A. Kalymon, "Economic Incentives in OPEC Oil Pricing Policy," _Journal of Development Economics_ vol. 2, 1975, pp. 357-362; S. D. Krasner, "Oil is the Exception," _Foreign Policy_ no. 14, 1974, pp. 68-84; Paul Leo Eckbo, "The International Petroleum Market: Some Behavioral Options," Ph.D. thesis, MIT, September 1975.

market will not cause tension among the members because most of the largest potential exporters do not "need" the revenues--is doubtless justified for some level of earnings from oil exports. Above some threshold of petroleum income, Saudi Arabia, Abu Dhabi, Kuwait, Libya--even Iran or Iraq--will be indifferent as to whether they have to forego some added increment of revenue. But "savers" are also "spenders" in OPEC--within some range lower current revenues begin to "hurt". That is to say, within some range of government revenues a hundred million dollars less in petroleum earnings means giving up some item--a desalination plant, a fertilizer project, a squadron of fighters, a promise of aid, a program of food subsidies to fight inflation and pre- serve domestic tranquility--that constitutes a "sacrifice." Where is this range? Where does the marginal utility curve for additional revenues begin to turn up?

A schematic representation of this problem looks like the following:

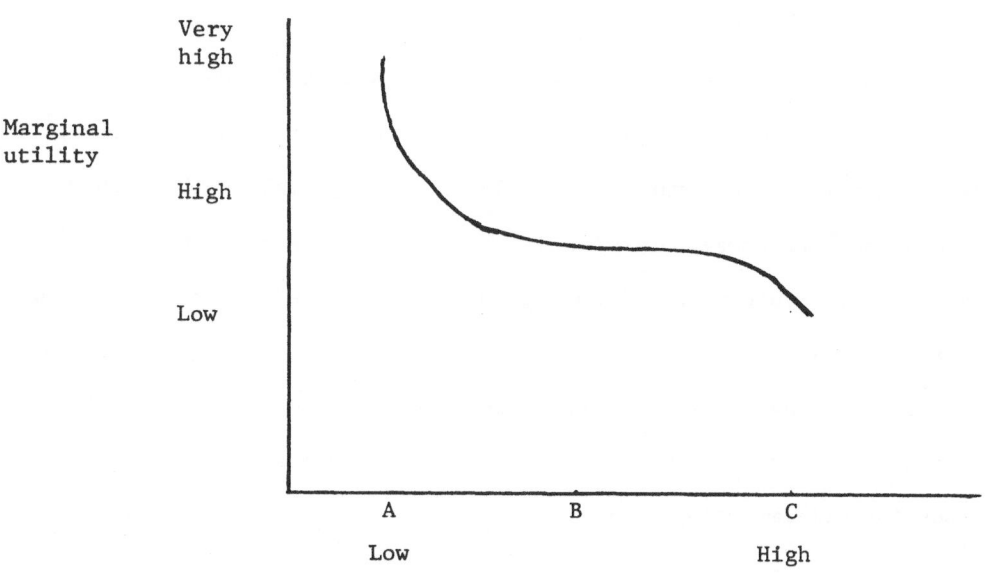

Energy revenues

Between points B and C, further energy revenues begin to become a matter of indifference to those domestic groups that preside over the implementation of social, economic, military, or international programs. The ability to absorb spare export capacity and forego the revenues in the interest of the common good of the cartel is relatively great. Between points B and A, important domestic groups begin to cross a threshold of pain as they are required to limit or reduce their spending programs. Within this range of energy revenues, spare capacity, if available, is increasingly hard to maintain idle. A government here has four choices: 1) it can continue to sacrifice in the common interest despite the domestic pain; 2) it can insist that its cartel partners allow it a larger market share while absorbing more idle capacity themselves; 3) it can cheat on the rest of the cartel by shaving its price to expand its exports; or 4) it can propose that the cartel ease the tensions of allocating market shares and reduce the temptation for cheating by raising prices insofar as market conditions permit.

Where is this threshold of pain?

The first section of this study attempts to answer this question. It attempts to sketch the marginal utility curve for energy revenues for the OPEC countries, especially Saudi Arabia, Iran, Iraq, and Venezuela. The second section, then, matches these estimates of the aggregate revenue "needs" against alternative projections of demand for OPEC exports, thus establishing how easy or difficult the "balancing problem" will be for OPEC at various oil price levels. The final section suggests which hypotheses about OPEC behavior are more, and less, plausible for the future, and discusses the policy implications for oil-importing states.

II. THE REVENUE "NEEDS" OF THE OPEC STATES

What is the shape of the marginal utility curve for energy revenues for the members of the Organization of Petroleum Exporting Countries? This is not primarily an economic question--it is a political question, "political" in the broadest sense of the word. It asks: at what level do additional revenues from oil (or refined products, or natural gas) become superfluous for those countries; at what level does a decline in energy revenues constitute a "squeeze" on the budgets of the countries; at what level does a further reduction in energy revenues mean that the government must "sacrifice" some program that political groups concerned with domestic stability, or national defense, or economic development, or international influence consider "important"?

To calculate the opportunity cost of foregoing some hypothetical increment of public revenue for Saudi Arabia, Iran, Iraq, and Venezuela, this chapter tries to estimate how far the governments of these countries will have to cut back on their announced social, economic, or military programs at a given production or price level for oil. It begins with the Development Plans of the four countries as the most comprehensive statement of national goals and priorities. But, as will be seen, the Plans alone pose serious problems as the basis for such calculations: first, they substantially underestimate the costs for the projects contained in them; second, it is unclear to what extent the Plans themselves are, or ever were, accurate reflections of the aspirations of the most important political groups in the countries under consideration; third, there are needs and demands that have appeared since the publication of the Plans that are not adequately reflected in the Plan documents. The Plans should be viewed, therefore, as only the entrée to a more elaborate, and more complicated, attempt at social analysis. For Kuwait, Qatar, the UAE, and Libya the problem is more complex still, since the ability to affect world events, to influence intra-Arab politics, to maintain a balance of power and prestige among themselves is more important in determining their energy export preferences than the level of domestic spending. For these countries, this study places more emphasis upon where in the past they have protested the decline in their exports and begun to discount their own oil so as to regain their market shares as an indication of the shape of their revenue utility curves. For the remaining high-population, high-mobilization members of OPEC (Nigeria, Indonesia, Algeria, Ecuador, and Gabon) this chapter assumes that the governments will prefer to produce oil and natural gas at a rate as close to full capacity as possible. Needless to say, few of the OPEC governments will be able to do exactly what they "prefer." The purpose of this chapter is to provide some basis for estimating how far from their pref-erences how many of the members will be pushed in the effort to balance supply and demand at a given price.

This chapter attempts a task that is impossible to accomplish, in eco-nomic or political terms, with any great degree of precision. The social value attached to programs enumerated in the spending plans of the governments, the priorities among announced objectives, and the estimates of the future costs are all too uncertain to permit any but the most broad-brush calculation of the marginal utility curve for energy revenues. Nevertheless, if the proper cautions are observed by the reader and proper precautions are taken by the analyst, it is possible, with some degree of confidence, to construct a frame-work for predicting at what revenue levels the problem of apportioning market shares will be easy and when it will begin to increase in difficulty.

1. Saudi Arabia

The principal observation to be made about expenditure projections for the OPEC countries should come as no surprise, but some of its implications will: "development," "defense," "social welfare," and "international influence" are extraordinarily expensive goals. Even cutting back on development, in the hope of restoring a measure of the previous social stability (lower inflation, adequate water and sewage, less congestion), requires large outlays of money. Consequently, the revenue level at which an extra hundred million dollars from oil exports becomes a matter of indifference to OPEC governments is considerably higher than has frequently been assumed. The principal observation to be made about Saudi Arabia's revenue needs should likewise be a commonplace: Saudi Arabia is no exception. The authorities in Riyadh, like their counterparts elsewhere in OPEC, will have to reconcile themselves to a future in which they will have to spend much more to get much less than their planners calculated in the euphoria of the 1973/74 petrodollar surplus.

How much will Saudi Arabia have to spend in the years ahead to accomplish portions of the Second Five Year Plan that elites concerned with social welfare, domestic stability, national defense, or economic development consider important? To make such estimates, one cannot use the figures given in the Plan directly for three reasons: first, many (if not most) of the projects in the Plan did not have adequate feasibility studies to use as the basis for their cost projections; second, the diseconomies of scale associated with rapid economic growth in an environment where there have been physical and administrative bottlenecks much more severe than originally anticipated mean that the Plan programs were severely underbudgeted; third, building and materials costs inside and outside of Saudi Arabia have escalated since the Plan estimates were made at a rate higher than the general OECD (or worldwide) price index, meaning that the "constant price" assumptions of the Plan are untenable.

The Second Five Year Plan, 1975-80, submitted to King Khalid in April 1975 was drawn up under very tight time constraints and with limited staff support. Various ministries relied for their cost projections upon earlier appraisals for programs uncompleted in the First Plan, upon rough international price comparisons for new projects, and upon some contemporary feasibility studies.[1] The most careful of the latter were carried out within the Ministry of Petroleum and Mineral Resources and the Ministry of Commerce with the help of foreign consultants (primarily Arthur D. Little), but even these were done in a great hurry. The projections for the Ministry of Defense and Aviation, in contrast, included no systematic attempt to relate the cost of specific weapons systems, or military construction projects, to the figures given in the Second Plan. For those projects for which feasibility studies were supplied, Saudi planners employed a 25-35 percent "locational premium" to account for the difference between construction costs in the Arabian Gulf and those in the home country of the bidders (preeminently the United States). In addition to the necessity of presenting cost projections without time for adequate pre-construction estimates, the Plan represented all programs as if they would be included in their entirety in the first year's budget (1395-1396 = 1975/76), with no diseconomies of scale. Finally, the Second Plan consciously left to one side the question of the impact of foreign or domestic inflation on project costs and attempted to represent all expenditures associated with the Plan in constant 1974 dollars.[2]

The Saudi Second Plan was drawn up in the middle (1974) of three inflationary explosions. First, OECD inflation amounted to about 29 percent from 1973 to 1976 (average increase in GNP deflators, measured in national

---

[1]This analysis of the Second Five Year Plan draws upon interviews with senior officials in various ministries of the Saudi government, especially in the Central Planning Organization, with U.S. government officials having responsibility for Saudi affairs, and with U.S. companies and consulting firms closely associated with the Saudi planning process, March 1976-May 1977.

[2]Given the time constraints and uncertainties that were present when the Second Five Year Plan was drawn up, the methodology of the Central Planning Organization may well have been the most sensible approach at that time.

currencies). Second, as we shall see, heavy construction and machinery costs for the kind of major projects contemplated for Saudi Arabia rose even faster than the general OECD price index (e.g., 65-90 percent for chemical plants, refineries, and other industrial construction in the United States, 1973-76). Third, domestic inflation in Saudi Arabia rose from 4 percent in 1972 to 16 percent in 1973, 21 percent in 1974, 36 percent in 1975 and 1976, with certain key components (e.g., industrial wages) rising much faster.[3]

Due to the fact that Saudi planners keyed the expenditure projections of the Plan to U.S. cost estimates plus a 25-35 percent "locational premium" for Saudi Arabia, it is not difficult to recalculate what the programs included in the Plan would really cost in 1977 prices. (Note: two "recalculations" are being combined here: a recomputation of the "true costs" of completing specific projects in Saudi Arabia and a transference from 1974 to 1977 prices.) One needs only an index of how fast construction costs of the kind envisioned for Saudi Arabia have escalated in the United States between 1974 and 1977, and a new calculation of the actual "locational premium" needed to undertake a project in Saudi Arabia in the current period. One can then multiply the Plan estimate for any given capital expenditure by the escalation in U.S. costs from the time of the original estimate to the present. This produces a new "U.S. base price" for the project. The new "U.S. base price" can then be multiplied by the difference between the current locational premium and the original assumed locational premium to produce an up-to-date estimate of the project's capital cost.

---

[3]U.S. government estimates. The "Economic Trends Report: Saudi Arabia" issued by the U.S. Embassy in Jiddah indicates that "the official price index indicates a 30 percent annual rate of increase (through the first six months of 1976), but the index either understates or ignores the most explosive elements of the current inflation--major consumption of capital goods, imports, housing and labor, to name the most important." (December 1976, p. 4.)

To calculate how much heavy construction costs have risen in the United States between the time Saudi program costs were first estimated and mid-1976, I have relied upon two public indexes of construction costs (the Chemical Week index of plant construction, and the Nelson oil refinery inflation index) plus the proprietary indexes of three large international construction companies. On the high side, Chemical Week reports that plant construction rose 40-56 percent in 1974, and 11-18 percent in 1975, and an estimate of 12-13 percent in 1976, or about 37-46 percent from mid-1974 through 1976.[4] On the low side, the Nelson refinery inflation index rose 14 percent from mid-1974 to 1976.[5] Nelson reports, however, that the "actual cost inflation" was greater than his index indicates due to the need to pay premiums that his figures do not record. The three proprietary indexes are somewhere in-between, with general construction costs rising between 25-35 percent mid-1974 to mid-1977. This study assumes that U.S.-based estimates for construction costs, therefore, would have to be raised by at least 30 percent to bring the figures from mid-1974 to mid-1977 prices.

To try to determine how much of a "locational premium" Saudi authorities must currently accept as a basis for awarding bids in Saudi Arabia (for sites in Jiddah, Riyadh, Damman, Jubail, and Yanbu), I have relied upon a survey of more than fourteen U.S. construction and operating companies or consultant groups plus the U.S. Army Corps of Engineers to try to determine how much more projects will cost in easily accessible Saudi sites as compared with the Texas Gulf Coast.[6] (In the case of some of the larger and more

---

[4]Chemical Week, December 17, 1975, p. 37; cf. also, Chemical Week, November 12, 1975.

[5]The Oil and Gas Journal, January 26, 1976, and May 3, 1976.

[6]Interviews conducted in the United States and Saudi Arabia, March-August 1976.

experienced companies—Bechtel, Fluor, Aramco, Exxon, Mobil—I have
conducted multiple interviews, both in the United States and in Saudi Arabia,
covering different types of construction. In the case of military construction
I have relied on the results of successful bids selected from a variety of
U.S. companies via the U.S. Army Corps of Engineers.) This "locational
premium" includes transportation costs, delays due to bottlenecks in Saudi
Arabia, service fees, and mini-infrastructure expenditures during construction (housing, sewage, power, labor procurement). It also takes into
account design modifications (extra capacity, back-up facilities, spare equipment) that are necessary to ensure quality comparable to a U.S. location.
The latter point is important since construction companies (e.g., Bechtel)
typically report locational premiums lower than those used in this analysis.
Their standard of comparison is the cost differential between comparably de-
signed plants in the United States and Saudi Arabia. But operating companies
(e.g., Exxon) report that comparable facilities will not give comparable quality of performance. Thus, Saudi plants must include more redundancy, more
in-house repair capability, more spares than would normally be required in
Galveston. The locational premiums used in this analysis are based on comparability of performance rather than comparability of design. The locational premiums indicated here do NOT, however, include demurrage charges.
Finally, of course, they include the impact of local inflation on the domestic component of the Plan projects as of mid-1976. I have then adjusted
these premiums to reflect mid-1977 prices.

The results of the survey suggest that the following "locational premiums" were appropriate as of mid-1976.

1.   Industrial construction = 2.1 x Texas Gulf Coast costs.

2.   Electrical power generation = 2.3 x Texas Gulf Coast costs.

3.   Housing, apartments, office buildings = 2.4 x Texas Gulf Coast costs (although some hotel construction is estimated at 2.8 x New York City building costs, and some simple concrete block housing has cost 3.5 x Texas prices).

4.   Hospitals = 2.5 Texas Gulf Coast costs·

5.   Military construction = 2.8 x Texas Gulf Coast costs.

Project costs as reflected in bids actually tendered have in fact escalated much more rapidly than the above premiums would indicate. The price of a large telecommunications project, for example, rose 600 percent (from $1.1 billion to $6.8 billion) while a recalculation of the locational premium (from 0.25 to 2.3) would predict a jump of "only" 80 percent. This has been due, in large part, to the Saudi insistence on fixed price contracts that forced the companies of necessity to include a large component for inflation over the life of the project. In early 1977 the Saudi government began to allow inflation escalator clauses in contracts while vowing to take a tougher look at locational premiums.[7]   This study combines the locational premiums indicated above with an assumption that the Saudi authorities will be able to bring the local inflation rate to OECD levels by 1980. This assumption may well be too optimistic. (See pp. 15,74).

---

[7]The telecommunications project was one that caused enormous controversy in Saudi Arabia. Cf. "Saudi Arabia Asks New Bids by Firms for Phone Project." Wall Street Journal, March 30, 1977.

Numerous other assumptions are necessary to estimate how much portions of the Plan will really cost. They are given in detail in the appendix on Saudi Arabia. Only two are of sufficient importance to mention here: the costs of "administration" for the Saudi government, and the "recurrent costs" associated with Plan projects. Administration: administrative costs amounted to more than $5 billion in the first year of implementation for the Second Plan (1975/76), in comparison with a Plan projection of $1.8 billion. Some of this first year's expenditures were doubtless extraordinary start-up costs, but most evidence indicates that the Saudi government is still very shorthanded in terms of trained administrators and modern administrative machinery (from office buildings to computers). Thus, it is difficult to imagine that larger proportions of the Plan will be administered with a smaller budget in the future than in the past. Nevertheless, to be cautious in making projections, this study assumes that 25 percent of the 1975/76 expenditures can be "squeezed out" of future budgets. "Recurrent costs" include the cost of keeping existing facilities operating plus local operating costs of new projects as they come on-line. This study assumes that 10 percent of the recurrent costs in the Second Plan can be attributed to programs inherited from the First Plan (since the ratio of the First to the Second Plan is about 1 to 10). The remaining 90 percent of the recurrent costs in the Plan are prorated, by category, depending upon what proportion of the Plan's projects are assumed to be in process.

For purposes of comparison with the original Plan, it is possible to calculate what the true cost of the Plan would be in 1977 costs. This calculation is given in table 1. But the reader should note that the figures given far understate what would in fact have to be spent to complete the Plan since it implicitly assumes that everything is accomplished at once with no further diseconomies of scale in implementation. This is plainly unrealistic.

Table 1

REVISED ESTIMATE OF SAUDI ARABIA'S SECOND PLAN COSTS IN 1977 PRICES

		Original plan		1977 revisions
		billion Saudi rials	billion U.S. dollars	billion U.S. dollars
I. Economic Resources (less original gas gathering and de- salination estimates)	recurrent project	2.0 39.4	0.6 11.2	0.9 26.5
1977 desalination estimate				18.0
1977 gas gathering estimate				16.0
II. Human Resources	recurrent project	43.9 36.2	12.5 10.3	19.2 28.4
III. Social Development	recurrent project	18.1 15.1	5.1 4.3	7.9 12.4
IV. Physical Infrastructure	recurrent project	12.5 100.4	3.6 28.5	5.6 71.1
V. Administration				
75% of $5.8 billion = $4.4 billion x 5 yrs.				22.0
VI. Defense	recurrent project	14.7 63.5	4.2 18.0	6.4 42.5
VII. Other	recurrent project	54.9 8.6	15.6 2.4	20.2 6.6
Total				296.5

Notes: 1) Conversion: 3.52 rials per dollar.

2) This calculation assumes that there are no diseconomies of scale in implementation that are not already included in the revised locational premiums.

3) The original plan totaled $144 billion, including the initial estimates for desalination and gas gathering facilities.

More useful than trying to demonstrate what the whole Plan would cost if it were done all at once is to calculate what various portions will cost as the Saudis in fact cut back and stretch out the programs. To do this, it is necessary to multiply any portion of any project to be completed after 1977 by the amount Saudi inflation for the local expenditure component of the project exceeds the OECD rate of inflation between mid-1977 and the date of the expenditure. The inclusion of this "Saudi inflation premium" will give the cost, e.g., of a desalination plant scheduled for 1980, in 1977 dollars. To approximate this "Saudi inflation premium," I have taken an estimate far more optimistic than any Saudi officials or expatriate advisers were willing to make. The cost of living index in Saudi Arabia rose about 5 percent in 1971, 4 percent in 1972, 16 percent in 1973, 21 percent in 1974, and 36 percent in 1975 and 1976.[8] Some components, e.g., industrial wages and housing, rose much faster than the general price index. The "Saudi inflation premium" used for the projections in this study is 31 percent in 1976/77; 20 percent in 1977/78; 10 percent in 1978/79; 5 percent in 1979/80; 0 percent in 1980/81. Thus Saudi inflation rates are assumed (probably unrealistically) to be brought down to OECD rates by 1980. The domestic inflation rate is of course not independent of the level of government expenditures chosen by Saudi authorities, although increased public spending on some kinds of imports brings down the local inflation rate. Project costs in the Plan are assumed to include 50 percent local expenditures (including expatriate labor whose

---

[8]These are U.S. government estimates. Private calculations suggest that Saudi inflation has been running (1975/76) at a rate of 60 percent per year. Professor Samir Ahmed, lecture on the Saudi Arabian economy, "Energy and the Middle East," International Management and Development Institute, Johns Hopkins School of Advanced International Studies, Washington, D.C., December 1, 1976.

cost is calculated to rise at least as fast as the domestic rate of inflation). Equipment and material imports are assumed to rise in price no faster than the overall OECD price index, although many observers suggest that heavy machinery and military hardware are likely to have their own "inflation premium." Recurrent costs are assumed to be affected in their entirety by the "Saudi inflation premium."

To calculate what a given proportion of the Plan projects would cost on an annual basis in a given year (e.g., how much Saudi Arabia would have to have to spend in 1979/80 if it set a goal of trying to accomplish 25 percent of the Human Resource projects included in the Plan in five years), this study (1) multiplies the 1977 revised estimate for the category of expenditure by the target percentage to be completed; (2) divides the product from (1) by the target number of years for completion; (3) multiplies the local expenditures for each annual segment after 1977 by the cumulative "Saudi inflation premium" to that year. Based upon the deceleration of domestic inflation hypothesized earlier, the multiplier for 1967/77 expenditures is 1.31; for 1977/78 expenditures, 1.56; for 1978/79, 1.72; for 1979/80, 1.80; and for each subsequent year, 1.80. (One should recall that the projection of the "Saudi inflation premium" assumed here is widely believed by Saudi officials and expatriate experts to be too low.)[9]

I have then attempted two simulations, with an analysis of the impact on Saudi policy makers which each would entail. The first (table 2), an "Immediate Drastic Cutback" scenario, represents a siutation in which Saudi authorities

---

[9] Aramco, for example, estimates that despite Saudi concerns about inflation the company's own composite costs will be rising at a rate of about 15 percent per year through 1980, for an "inflation premium" of 9-10 percent above the OECD rate (December 1976).

Table 2

SIMULATION OF AN "IMMEDIATE, DRASTIC" CUTBACK
IN SAUDI ARABIA'S PLAN TARGETS TO 1979/80
(constant 1977 dollars)

		Total cost ($ billion)	Percentage cutback	1979/80 expenditures ($ billion)
I. Economic Resources (less original gas gathering desalination estimates)	recurrent	0.9	90	0.1
	project	26.5	90	0.7
1977 desalination estimate		18.0	*	1.8
gas gathering estimate		16.0	*	1.6
II. Human Resources	recurrent	19.2	75	2.3
	project	28.4	75	2.1
III. Social Development	recurrent	7.9	75	0.9
	project	12.4	75	0.8
IV. Physical Infrastructure	recurrent	5.6	75	0.7
	project	71.1	75	5.0
V. Administration		22.0	25	7.6
VI. Defense	recurrent	6.4	75	0.7
	project	42.5	75	2.9
VII. Other	recurrent	20.2	0	5.8
	project	6.6	0	2.0
Total				35.0

* stretched out over ten years instead of five.

(A "cutback" of 75 percent means that in any given year, e.g., 1979/80, only 25 percent of the projects contemplated in the Plan will be in the process of being implemented.)

in 1977/78 radically downgrade the Plan targets (including cutbacks in some categories from levels of spending already achieved, e.g., in Defense and Administration). It hypothesizes that the Saudi hierarchy considers domestic stability, on the one hand, and an ability to influence intra-Arab politics, on the other, to be of high priority, national defense and regional stability to be of middle priority, and economic development (especially import-substitution development that requires the importation of foreign workers) to be of much lower priority. As explained earlier, it is assumed that "Administrative" costs cannot be reduced below 75 percent of 1975/76 expenditures to oversee programs whose annual cost is in excess of $20 billion. Eighty percent of the Administration costs are assumed to be subject to the Saudi inflation premium. In addition, this simulation assumes that there will be growing pressure to increase the proportion of funds in the "other" category to be spent on food subsidies to ensure domestic tranquility and hold down local inflation.

The foregoing simulation represents a world in which Saudi officials are willing to cut general economic development projects by 90 percent (in comparison to the original Plan), defense spending by 75 percent, infrastructure, desalination, and gas gathering projects by 50-75 percent, educational, medical, and housing programs by 75 percent, while leaving foreign aid at its current level and increasing food subsidies (to help control local inflation) slightly. Another way of looking at this simulation is that it forces Saudi authorities to stretch their Plan from five to more than twenty years.[10] Even so, with constant real oil prices and a Saudi government revenue of

---

[10]The expenditure level in this scenario, like the one in the much larger simulation that follows infra, is considerably below the physical limitations on Saudi Arabia's capacity to import. For the discussion of physical constraints due to port, road, and air cargo limitations, see p. 24-25 and the Appendix on Saudi Arabia.

approximately $11.56 per barrel (1977 dollars), this "immediate, drastic

cutback" scenario will require exports of 8-9 million barrels per day (mbd)

to finance year by year, or 7-8 mbd by the late 1970s if the Saudis are

willing to draw down their financial reserves to zero in ten years. (As

shown in the Appendix, the level of "real, effective" Saudi official assets

will probably peak at about $57 billion in 1978.[11] Note: Saudi official

investment income is set at zero--a figure probably too high since Saudi

monetary authorities have not been able thus far to equal the OECD inflation

rate in their return on official assets.)[12]

How much of a "sacrifice" would this scenario mean for the various

groups emerging in the Saudi hierarchy and the various factions within the

royal family? The answer is impossible to know with exactitude, but one can

calculate what kind of hard choices the "immediate, drastic cutback" would

force upon the Council of Ministers. It would require the Council to make

decisions with regard to economic development such as: dropping entirely

all plans for a steel or aluminum industry and for local manufacturing

(automobile assembly or parts, health products, consumer goods, or food

processing); stretching out the timetable for development in the west along

the Red Sea so that the petrochemical complex and export refinery, fed by

the crude line and the NGL line, will not come onstream until sometime after

---

[11]As explained in the Appendix on Saudi Arabia, p. 89, "real" or
"effective" assets are those that the Saudi government could in fact draw
upon to finance its own budgetary needs. These include gold, hard currencies,
and stocks and bonds held by SAMA, and are estimated by the U.S. government
to have totaled $52 billion at end-1976. Other common estimates of Saudi
"official"assets" (e.g., by Morgan Guaranty) include a portfolio of loans to
other Arab states (Egypt, the Sudan, Jordan, Syria) that are unlikely to be
able to be called in to finance Saudi internal needs.

[12]Former Ambassador James Akins has reported that Saudi officials told
him that they had not been able to earn a rate of return on their financial
resources as high as the OECD rate of inflation. Thus they could not, and
did not expect to generate any real income from these assets. Rather, their
goal was to minimize what would probably be a steady real loss.

1990; and postponing two of three petrochemical plants, both fertilizer plants, the lube oil refinery and one of two export refineries until some-time after 1982.[13] In the social field, it would require Saudi authorities to choose in the next decade, for example, _either_ to double boys' education at the primary and secondary level (as planned) _or_ to proceed with the new Mecca campus for the King Abd Al-Azi University; _either_ to expand the provision for women's education _or_ to build the college of medicine at Jidda; _either_ to construct 212 small town and rural dispensaries _or_ to pro-vide 11,500 hospital beds for the kingdom. It would force Saudi military strategists to settle for what their American military advisers are telling them is the minimum for a "paper deterrence" (based, for example, around the F-5e versus Iran's F-14s and F-16s and Iraq's MIG-21s and MIG-23s) to ensure the stability of the Gulf in the _late_ 1980s.[14] In practice, of course, the Saudis will probably spread their cutbacks across all of their programs--social, economic, military, and international--rather than making dramatic either/or choices. But the cutbacks in this simulation will have to be of sufficient magnitude to make broad sectors of the Saudi hierarchy feel a tight fiscal pinch.

---

[13]While the hypothesized indifference of Saudi authorities to economic development objectives represented in this simulation seems highly plausible, several commentators have suggested that this scenario would exacerbate regional tensions in Saudi Arabia by delaying development along the Red Sea (where the merchant class has traditionally been most strong) until after 1990. It would also cause concern to Saudi defense strategists who are anxious to construct Red Sea petroleum loading facilities as an alternative to total dependence on passage of tankers through the Gulf. Yet to proceed with substantial west coast development before 1990 requires immediate action on the cross-peninsula pipelines and related infrastructure. Thus the low priority accorded economic development expenditures in this simulation has been considered by several commentators to be unrealistic.

[14]These figures do not include a possible military "inventory" maintained for the "frontline" Arab nations. Such purchases would have to be added to the outlays calculated here.

The discussion of any kind of cutback on Plan targets whatsoever has
been extraordinarily muffled. When Dr. Ghazi Abd al-Rahman al-Qusaibi,
Minister of Industry and Electricity, spoke in Washington in 1976 about the
possibility of cutting down and stretching out the Plan, he was made to
retract his statement publicly when he returned to Riyadh. In Saudi Arabia,
the budget requests from the individual ministries were 60 percent higher
for 1977 than they had been for 1976, despite the fact that only about 60
percent of the 1976 appropriations were actually spent, except in the areas
of Administration and Defense (where the expenditure rate was much higher).[15]
These enormous requests from the ministries, plus concern about the high
rate of inflation, forced a confrontation with the Ministry of Planning over
the 1977 budget. "It is politically impossible to tell the Ministries to go
slow, establish priorities, conduct feasibility studies," Dr. Faisal Bachir,
Deputy Minister of Planning for the National Economy, has lamented, "when
the Ministers are convinced that there are unlimited funds."[16] The Planning
Ministry emerged "victorious," however, with a budget for 1977 "held" at the
1976 ceiling of $31.4 billion. For 1977, in contrast to earlier years, most
observers felt that all of the funds appropriated would be spent, and then
some.[17]

---

[15]Estimate by the Saudi Arabian Monetary Authority (SAMA), January 1977.

[16]Interview, Riyadh, June 26, 1976.

[17]Late in 1977 the Saudi Arabian monetary authority announced that actual
expenditures for fiscal year 1976 (ending June 15, 1977) were higher than the
"nominal" budget ceiling of $31.4 billion. For fiscal year 1977 (ending
June 15, 1978) wage outlays were scheduled to be double the level of the previous year. Even with no real growth in project spending, or in international
aid, the government would probably have to spend more than $36 billion (or
$3 billion more than the "momentum and crunch" scenario simulated infra, sug-
gesting that the scenario may underestimate actual expenditure patterns).

By 1977 the major infrastructure projects, gas gathering, desalination, military construction, and industrial parks (at Yanbu and Jubail) had only barely been begun. Even if, as expected, they were cut back and stretched out, they would generate an enormous momentum in public expenditure.

In the light of this spending momentum, a second cutback scenario (a "momentum and crunch" simulation) has been constructed (table 3). It represents a world in which the current process of letting contracts for physical infrastructure, for military facilities under the direction of the U.S. Army Corps of Engineers, for urban construction in Jidda and Riyadh, and for the beginning of industrial development at Jubail and Yanbu continues through 1977/78, while health and educational expenditures, which have been slow to materialize in 1976/77 grow steadily to a level where 25 percent of the projects envisioned in the Second Plan are under way by 1980. More specifically, it assumes that (1) the negotiations with Aramco (and subcontracts with Fluor, etc.) for the gas gathering system proceed, although the entire project will be stretched out over ten years, rather than the five-year period contemplated in the Plan; (2) defense spending continues at the 1976 rate, paced by the Corps of Engineers' supervision of military construction whose current inventory of Plan projects will be completed by 1982, only two years late; (3) infrastructure projects (especially port and internal transportation) continue to receive relatively high priority, with the goal of having half of the Plan programs in progress by 1980 and the entire Plan completed by 1990; (4) other economic resource contracts are comparatively de-emphasized, with only 25 percent of the desalination projects and 10 percent of the other development projects scheduled to be begun before 1980; (5) education (Human Resource) and health (Social Development) programs that appear to have long start-up times grow to a level of 25 percent of the Plan by 1978/79.

(Note: The reader should have more confidence in the expenditure projections at the end of this period, if current spending momentum continues, than in the exact path that annual outlays take in the interim.)

Table 3

"MOMENTUM AND CRUNCH" SCENARIO FOR SAUDI EXPENDITURES, 1977-80
(1977 prices in billion dollars)

		1977/78	1978/79	1979/80
I. Economic Resources				
(less original gas	recurrent	0.1	0.1	0.1
and desalination)	project	0.6	0.7	0.7
1977 desalination		1.8	1.8	1.8
1977 gas gathering		1.6	1.6	1.6
II. Human Resources				
	recurrent	0.7	0.7	0.8
	project	0.6	1.2	2.1
III. Social Development				
	recurrent	0.4	0.7	0.9
	project	0.3	0.5	0.8
IV. Physical Infrastructure				
	recurrent	0.5	0.8	1.2
	project	3.8	6.8	10.0
V. Administration		6.7	7.4	7.6
VI. Defense				
	recurrent	1.5	1.7	1.7
	project	7.8	8.3	8.5
VII. Other				
	recurrent	5.1	5.4	5.6
	project	1.9	2.0	2.1
Total		33.4	39.7	45.5

This scenario would require 10.8 mbd in oil exports to finance on a current basis in 1979/80, or approximately 9.8 mbd if the Saudis are willing to use about $4 billion from the country's financial reserves in 1980 (a rate that would deplete them in ten years). (See pp. 88-90.)

This "momentum and crunch" scenario would lead the Saudis more gently into the recognition of financial as well as physical constraints on the transformation of their society than the "immediate, drastic cutback" simulation. But the bottom line does not disappear. Saudi authorities might well order their spending priorities differently and shift resources from social programs, for example, to economic development. The basic financial constraints remain. Even trying to "cool" the economy--reduce inflation, reduce congestion, reduce confusion--requires enormously costly programs to expand import capacity, create infrastructure, renovate utilities, subsidize food. To finance these expenditures, Saudi authorities will, by 1978, either have to expand exports rapidly at current prices, or press for higher prices, or draw down on their financial reserves, or begin to cut back on infrastructure development and on military contracts already awarded, while letting no major new contracts in these areas, in order to make room for relatively low but growing levels of social welfare (housing, medical and educational). The need to cease awarding many new contracts in 1978 will hit particularly hard the many members of the Saudi hierarchy whose family income depends upon commissions from the successful negotiation of terms for new projects.

Both of these expenditure patterns--including the $46 billion "momentum and crunch" scenario--are well within Saudi Arabia's physical capacity to import. U.S. government studies indicate that Saudi Arabian "effective" import capacity (ports, overland, air freight plus warehousing and internal distribution capabilities) should rise from about 7,795 thousand tons per year in 1976 to 15,800 thousand tons per year in 1980, an increase of about 100 percent.[18] If one uses 1975 unit import values per ton for cement, grain, and general cargo, and assumes a roughly equivalent composition of imports in 1980 as in 1975, Saudi tonnage capacity can be converted to an import value (in constant 1975 dollars) of $18.7 billion. The "momentum and crunch" budget for 1980 includes $17.9 billion in current expenditures, administration, and foreign aid whose aggregate import component will be about 10 percent, and $27.6 billion in project expenditures whose import component is estimated at 30 percent. This produces an import total associated with the government budget of $10.1 billion. If the import of consumer goods and materials and manufactures for the private sector expand in line with a real non-oil GDP growth of 15 percent per year 1976-80, this will account for an additional $5.2 billion in imports. The result is a total import bill of $15.3 billion, with an import capacity of $18.7 billion. This leaves room for 500,000 foreign workers ranging from managers and professionals to skilled and semi-skilled workers (distributed in accordance with the projected manpower needs of the Plan, with a mean disposable income of $16,000 in 1977 dollars) to spend about 40 percent of their wages on direct imports each year ($3.2 billion). (Since the entire Saudi Plan estimates that no more than 500,000 foreign workers will be needed, and the "momentum and crunch" simulation assumes that only 25-50 percent of the Plan projects are in progress at any one time, the import calculation for the same number of workers should be more than adequate).

The conclusion is remarkable: the $46 billion "momentum and crunch" scenario clearly fits within Saudi physical import capacity, and anything less than that might require Saudi docks and warehouses to stand idle some of the time! Rather than let this happen, one supposes, Saudi authorities

---

[18]For details on Saudi physical import capacity, see the Appendix on Saudi Arabia, p. 90.

will increase the flow of imports to the private sector to ameliorate the rate of domestic inflation.

The two simulations presented here do not assume that Saudi Arabia "must" balance its fiscal budget on a current basis every year. The Saudi authorities may indeed draw down their financial reserves more rapidly than the ten-years-to-zero-rate hypothesized above. Or they may borrow abroad. Or they may tax foreign construction companies.[19] Or they may institute an income tax on Saudi citizens. The only argument advanced here is that below an annual revenue level of about \$35-\$46 billion in 1980 (1977 prices) the policy choices faced by Saudi authorities become increasingly painful.

The foregoing analysis suggests that the marginal utility curve for energy revenues for Saudi Arabia might look approximately like the following in the latter part of the 1970s and early 1980s (in 1977 dollars).

Saudi Arabia's "Need" for Energy Revenues in 1980

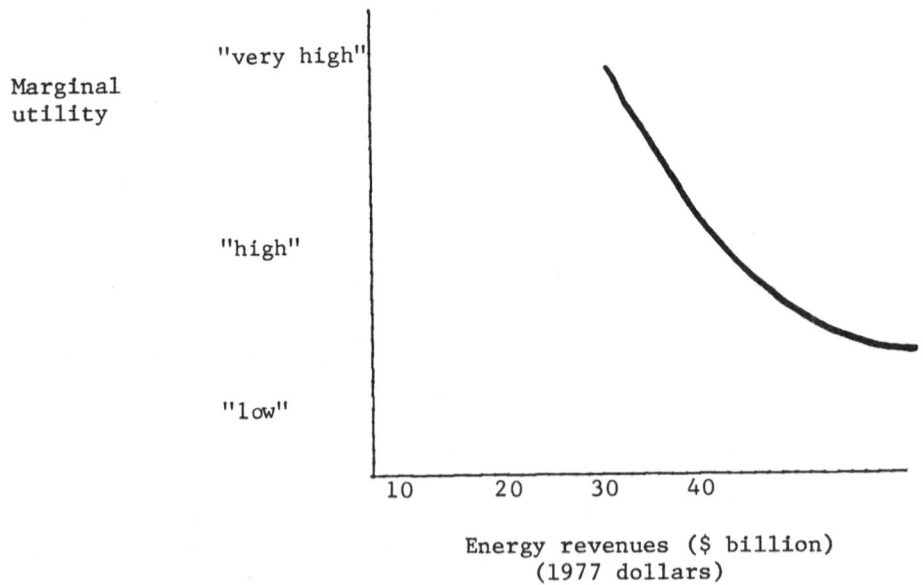

Energy revenues (\$ billion)
(1977 dollars)

---

[19]An effort to increase Saudi government revenues by increasing taxes on the foreign business sector will simply result in higher project costs, although the pass-through may of course not be one-for-one.

In short, for reasons of stability, welfare, development, defense, and influence, Saudi ministers and generals will consider it not only a "sacrifice" but a serious sacrifice if they are unable to finance programs at a rate of $35-$46 billion (1977 prices) in the early 1980s. How does this translate into a "preferred" level of oil exports? If one hypothesizes that real oil prices remain constant between now and then, and that the Saudi hierarchy is reluctant to draw steadily down on its financial reserves very fast, the Saudi government will prefer to export at least 9-10 mbd in 1980. Stated differently, exports of less than 9-10 mbd oil equivalent (petroleum, natural gas products, petrochemicals) will produce a fiscal "pinch" on the Saudis. This does not mean, of course, that they will have any inclination to break away from OPEC if their exports fall below 9-10 mbd. It simply means that they will probably insist that their colleagues--Iran, Iraq, Nigeria, Indonesia, etc.--help them bear the burden of balancing supply with demand. Or, if this proves too difficult, they will have to consider supporting a move for higher real prices within OPEC. In either case, one point is evident: soon Saudi Arabia can hardly be expected to balance supply and demand for OPEC by itself if that means its market share must drop much below 9-10 mbd.

But what about those in the royal family or in the Council of Ministers who dislike the high level of economic activity and the pace of public spending in Saudi Arabia, and who would like to go back to the relative quiet of the pre-1973 period? Or what about the assertion, made by Sheik Yamani and others, that Saudi Arabia could cut production to 3 mbd, or even 1 mbd, without feeling the effects? Due to the high current level of official reserves ($49 billion at end-1976; see appendix, pp. 89-90), the Saudi government could easily do this for a short period of time. Thus the threat should be given serious consideration in the event of another Middle East war (although the behavior of other oil exporters with excess capacity at the time would be crucial to the effectiveness of a Saudi production cut). As a longer term strategy, however, one has to weigh the credibility of the Yamani assertion against the kind of domestic upheaval that might result against a future government that tried it. A Saudi government that in 1977 adopted the "immediate, drastic cutback" scenario presented above would, by 1979, have an annual budget of $27.7 billion remaining even if it suddenly reduced expenditures on infrastructure, gas gathering, and industrial projects to zero. That would require exports of 6.6 mbd to finance on a current basis. It could cut its budget to $25.4 billion by also stopping municipal water projects completely and eliminating the entire food subsidy program. That would require exports of 6.0 mbd. It could cut its budget to $21.6 billion by also firing half of the bureaucrats in the Saudi administration. That would still require exports of 5.1 mbd. It could cut its budget to $17.7 billion by also stopping all foreign aid (predominantly to other Arab states). That would require exports of 4.2 mbd. It could cut its budget to $14.8 billion by also eliminating all expenditures on military construction and equipment. That would get to the 3.5 mbd mark. And so on.

The record of survival for regimes that try to stop, or reverse, the process of social mobilization and rising expectations once it is started should not recommend this as a promising strategy for a monarchy.[20]

2. Iran

The marginal utility curve for revenues from the energy sector in Iran is much easier to draw than the corresponding curve for Saudi Arabia: It is

---

[20]Cf. Samuel P. Huntington, Political Order of Changing Societies (New Haven, Yale University Press, 1968) ch. 3,4; Mancur Olson, Jr., "Rapid Growth as a Destabilizing Force," Journal of Economic History vol. 23 (December 1963); James C. Davies, "Toward a Theory of Revolution," American Sociological Review vol 27 (February 1962); Karl W. Deutsch, "Social Mobilization and Political Development," American Political Science Review vol. 55 (September 1961).

high for the entire range of values approximating current prices times current export capacity, and does not begin to decline to any great extent until revenues increase substantially.

Like Saudi Arabia, physical rather than financial constraints were at first perceived as being the principal barrier to the achievement of rapid national development in the aftermath of the 1973 oil price jump.[21] The Fifth Plan, as presented in revised form in March 1975, envisioned the accomplishment of major programs in industrial development, defense, and social services totaling $112 billion with a net revenue surplus, after the completion of all projects, of more than $11 billion. There was widespread discussion of how Iran's monetary reserves could be invested abroad to produce the income needed to sustain domestic growth after the oil exports began to decline in the mid-1980s. But this euphoria dissipated rapidly. By the end of 1975, the Iranian government faced both balance-of-payments and fiscal deficits. This was due, in an immediate sense, to an 11 percent drop in demand for Iranian oil (from 6.0 mbd in 1974 to 5.4 mbd in both 1975 and 1976). But the more fundamental problem for Iran, as this section will demonstrate, is that the kind of social transformation needed to create even a middle-range industrial power is enormously expensive.

Unlike Saudi Arabia, however, Iranian planners have had an acute appreciation of the finiteness of their petroleum reserves. The basic production profile developed by the National Iranian Oil Company (NIOC) and by the Iranian Plan and Budget Organization (PBO) projects oil capacity rising from 1973 to 1978 (to about 7 mbd), then six years of steady output (to 1984),

---

[21]This analysis of the Iranian Fifth Plan draws upon interviews with responsible officials in various agencies and Ministries of the Iranian government, especially in the Plan and Budget Organization, with U.S. government officials in Washington and Tehran, and with U.S. companies and consulting firms familiar with the Iranian planning process, March 1976–May 1977.

followed by a steady and irreversible decline. Even among those planners who have criticized the wastefulness of the government's "all-out" strategy for development following 1973, the Fifth (1973-78) and the Sixth (1978-83) Plans represent the crucial period of a "big push" to transform Iran into a major power with a diversified industrial and primary export base.

How closely do the figures given in the Fifth Plan represent contemporary program costs? The original Fifth Plan was drawn up in 1972/73 (1351 in the old Iranian calendar, 2331 in the new Monarchic calendar) with an implicit assumption in the minds of the planners that project costs would rise 25 percent (3 percent per year) over the five-year life of the Plan.[22] The Fifth Plan was revised in 1974/75 by adding new projects to the original Plan and hiking the implicit assumed inflation rate to 12 percent per year. Thus, Iranian planners made some allowance for inflation that the Saudi planners did not. They also may have estimated the costs of the programs included in the Fifth Plan more accurately than did the Saudis. Finally, the Iranian economy is sufficiently large and complex as to prevent some of the extreme bottlenecks encountered on the other side of the Gulf. For these reasons, the recalculation of the cost of Iranian projects to include the actual rise in foreign and domestic prices does not present results as dramatic as in the case of Saudi Arabia. Nevertheless, the intensity of the financial squeeze that the country will continue to feel, even in the unlikely case that oil exports can be raised immediately to full capacity and held there indefinitely, as it tries to complete the Fifth Plan and move into the Sixth Plan, is clearly evident.

To gain some insight into the "need" that the Shah of Iran, the military services of Iran, various government agencies and social groups in Iran, and

---

[22]Firouz Vakil, Director, Planometrics and General Economy Bureau, Plan Organization, June 21 and August 24, 1976.

Iranian planners perceive the country has for revenues from the energy
sector, I have tried to estimate how many of the revised Fifth Plan projects
were accomplished by early 1976, and how much room a hypothetical Sixth
Plan (beginning in 1978) will have to accommodate the pressing new demands
for public sector expenditures into the 1980s.

To estimate how much progress has been achieved in accomplishing the
Plan targets, one must deflate the Plan capital expenditure figures by the
inflation rates implicitly assumed by the Iranian Plan and Budget Organization
(PBO), reflate the cost projections by more realistic calculations of the
rise in local project costs, and subtract from this total the actual budget
outlays in each category of the Plan each year. To accomplish this, this
section adopts the following methodology:

(1) it divides the original Fifth Plan public fixed investment figures
into five equal segments, and factors out an incremental 5 percent cost
increase in each of the years. This produces the capital cost estimates for the
the original Plan projects in 1351 (1972/73).

(2) it deflates capital expenditures for 1352 (1973/74) and 1353
(1974/75) by an index of the rise in construction costs between the time the
expenditure is made and the base calculation period (1351), and subtracts the
result from (1). This indicates the progress that was made in 1352 and 1353
toward completing the original Fifth Plan capital programs, and the amount
that remained as of the beginning of 1354 (1975/76) in 1351 prices.

(3) it divides the amount of new public capital expenditure in the
revised Fifth Plan (total public fixed investment in revised Fifth Plan minus
total public fixed investment in original Fifth Plan) into three equal segments (for the last three years of the Plan) and factors out an incremental
12 percent for each of those years. This gives the 1353 (1974/75) capital
cost estimates for the new projects added to the revised Plan.

(4) it reflates (2) and (3) by an index of the rise in construction costs

through (1975/76) since 1351 (1972/73) and 1353 (1974/75) respectively, and

subtracts from this the capital expenditures made in 1354 (1975/76). This

shows the progress made in the first three years of the Plan (to March 1976),

and provides a base for simulating future costs and future "progress."

The index of the rise in construction materials and construction wages

used here comes from the Iranian Plan and Budget Organization. The composite

index has been derived by assuming that roughly 40 percent of construction

costs are wages, as indicated in interviews with large U.S. companies

operating in Iran.

	Construction materials	Construction wages	Composite	Annual rate of increase
1351 (1972/73)	116.3	183.2	100.0	--
1352 (1973/74)	141.2	222.6	121.0	21%
1353 (1974/75)	180.2	286.8	151.3	25
1354 (1975/76)	188.0	414.4	181.6	20
1355 (1976/77)	NA	NA	NA	18 (hypothesis)

Whether the PBO index presented above adequately represents the increase

in capital outlays required for industrial, residential, military, and infra-

structural facilities between 1972 and the present, however, is open to some

question. A series of interviews with Iranian and American corporations and

analysts in Tehran suggested that these indexes may be too low, due to pre-

miums for materials and overtime (or fake overtime) in the tight Iranian labor

market. Some knowledgeable Iranians have argued, for example, that since 1974

labor costs in the construction industry have been doubling or tripling each

year. For an international comparison, the costs for capital investment in

the United States and Europe rose 90-115 percent for chemical plants between

1972 and 1976, 55 percent for refinery materials, and 50-90 percent for other

kinds of industrial and residential construction.[23] Thus, Iranian construction
costs are represented here as rising about in line with, or slightly less
than, price changes for similar projects in Europe and the United States.
This reinforces the suspicion that escalation for fixed capital investments
used here on the basis of PBO figures may be too low by a significant
amount.

Furthermore, it is assumed for the purposes of this exercise that all of
the original estimates of capital costs for economic, social, and public
affairs programs were accurate at the time they were made (1351 for the
original Fifth Plan, 1353 for the revised Fifth Plan). In Saudi Arabia, one
must remember, there was a systematic underestimation of original project
costs due to the diseconomies of scale (administrative and infrastructure
bottlenecks) when a large number of projects were initiated simultaneously.
Thus it was necessary in the Saudi case to allow for a one-time jump in the
base from which project costs were calculated in the 1973-75 period. One
might argue that there should be a strong presumption that there was a similar
process of systematic underestimation at work in Iran, especially for the
programs carried forward from 1972 before the enormous expansion of economic
activity began. But, due to the complexity of the Iranian economy and the
relatively small participation of foreign firms, it is much more difficult
in Iran than in Saudi Arabia to verify how contemporary costs compare to the
original estimates. The survey of more than 14 major international construction companies operating in the Persian Gulf indicated that project costs
in Iran ran from 1.3 to 2.5 times (or more) those for similar installations
in the Texas Gulf, depending upon the type of facility and the region of Iran.

---

[23]These calculations come from the sources in footnotes 4 and 5 plus
the proprietary indexes of three major international construction companies
doing business in the United States, Europe, and the Persian Gulf.

But it was not possible, in the Iranian case, to make simple generaliza-
tions about the extent to which accurate "locational premiums" were incor-
porated into the figures that appear in the Fifth Plan. The National Iranian
Oil Company (NIOC), for example, has tended to use a "locational premium" of
1.25 for bids accepted in the Abadan areas. Yet NIOC has its own cargo docks
and off-loading facilities (jealously guarded) that eliminate many bottlenecks
that exist for the rest of the economy, and that may make the 1.25 figure
reasonably accurate. Even where locational premiums can be shown to be high,
it is not clear that there are large cost overruns in comparison to the Plan.
The cost of the Dupont fibers facility, for example, is 50 percent higher
than what an identical facility would be in North Carolina, plus requiring
50 percent more working capital to cover spare materials and parts, but
almost all of these costs were anticipated in the original estimates.[24] There
are examples of small cost escalations: Pullman-Swindell, for example,
reports that costs in the Iranian steel industry have risen only 20 percent
in two years above what was originally estimated. There are examples of
larger cost escalations: the Sar Chesmeh copper mining facility, for example,
managed by Anaconda, has seen an escalation of 90 percent in four years above
what was contemplated at the outset. And there are examples of enormous cost
escalations: Medicorps, for example, won a bid for three hospitals in late
1974 and then reestimated the cost in August 1975 at twice the original
figure. The best way to represent the relationship between Plan estimates
and subsequent program expenditure requirements remains elusive.

To simplify matters, this study consciously errs on the side of caution:
it assumes explicitly that all original capital cost estimates for both the

---

[24] This and the following examples come from either the companies themselves, or from Iranian and U.S. government officials familiar with the case
histories. Interviews during March-August 1976.

original and the revised Fifth Plan were accurate at the time they were made with full knowledge that this assumption may systematically under-estimate Plan costs by a substantial amount.

In addition to the escalation in capital costs for economic, social, and military projects, a further item must be included to estimate progress made thus far on the Fifth Plan. Like Saudi Arabia, Iran has experienced a dramatic rise in the current account budget that is needed to maintain existing public programs and administer new ones. (For Saudi Arabia, these expenses come largely under the heading "Administration." For Iran, they come category by category under the heading "current" or "non-fixed" expenditures.) The revised Fifth Plan anticipated a ratio of current to capital costs of about 1 to 3 in the non-Defense categories (Economic Affairs/Social Affairs/Public Affairs). But in 1353 (1974/75) they were in fact almost 1 to 1 (0.97), and by 1354 (1975/76) had been lowered only slightly to 0.9.

If one squeezes the implicit PBO inflation assumptions out of the Fifth Plan and replaces them with more accurate calculations of the costs of Plan projects (for details see the appendix on Iran), one can see that as of the first quarter of 1976 the following progress had been made:

Percentage of Fifth Plan Capital Projects Completed (as of March 1976)[25]

Economic Affairs	23%
Social Affairs	30%
Public Affairs	36%

(Defense expenditures are not separated into capital and current accounts. As of March 1976, 83 percent of the Plan's total five-year projection for the Defense category had been spent. Furthermore, U.S. government analysts consider

---

[25]The data needed to extend the calculation to March 1977 were not yet available.

that at least 12 percent of the "social affairs" expenditures--70 percent of the outlays on "public housing," for example--are in fact military construction. In addition, 27 percent of the "public affairs" expenditures have gone to expand the government's multiple domestic security agencies.)

These figures suggest that Iranian planners have a long way to go to complete the programs envisioned in the Fifth Plan (which ends in 1978). The Central Bank of Iran estimates that demand for oil will run at an average annual rate of 4.6 mbd in 1977/78. If the government spends the revenues thus generated ($21.2 billion annually) in a pattern reflecting the priorities of the 1976/77 budget with a yearly deficit of $3.0 billion, the country will complete only 51 percent of its economic projects, 57 percent of its social projects, and 84 percent of its "public affairs" projects by the end of the Fifth Plan period.

Even if Iran were able to make the 15 percent price rise in 1977 stick and export at full capacity (6.8 mbd), the progress achieved would be 53 percent of the economic projects, 59 percent of the social projects, and 89 percent of the "public affairs" projects. In both cases about 12 percent of the social expenditures are non-civilian, and 27 percent of the "public affairs" expenditures are related to domestic security. Military expenditures will equal 103-116 percent of the Plan projection. Domestic demand for petroleum equals approximately 0.4 mbd per year. In each simulation, government revenues from non-energy sources are projected to grow at the PBO projection of 13 percent per year, and revenues from natural gas sales run at approximately $0.3 billion per year. In the first case, the Iranian government would have to finance an annual deficit of $3.0 billion with a combination of domestic and foreign borrowing.

It is clear, first, that there must be a large carryover of Fifth Plan projects into the so-called Sixth Plan period (1978-83), especially in the big

"economic development" category. Indeed, there will hardly be room for any
additional industrial projects in the next Plan if oil prices do not rise
in real terms which would be a disappointment to those who have seen the 1973-83
decade as crucial for economic "take-off" before the country's oil production potential dwindles. Second, it is evident that even (an improbable)
immediate expansion to full export capacity will not alter this situation a
great deal, especially in the short run. Increased demand for Iranian oil,
in the absence of further price rises, will not solve the country's
revenue squeeze problem.

One might argue that the "revenue squeeze" is, in some sense, a self-
delusion. The "big push" strategy for industrial development could be
stretched out over a longer period of time and financed by revenues accumulated
in the interim, or by oil "saved" in the ground to push back the day of de-
pletion.[26] Independent of the debate about what rate of government

---

[26]The Iranian case presents an interesting example of the problem of what
the impact of a high marginal utility for earnings from natural resource exploitation may be on the preferred rate of a particular government for exploiting
the resource. If the present and future prices of the resource, and the marginal costs of production, are known to the government, the economics of how
fast to exploit an exhaustible resource are fairly simple: the government
should convert the national patrimony into schools and jobs and military pro-
tection as long as the social rate of return from these expenditures is higher
than the appreciation of the resource in the ground. But this "simple"
analysis assumes that the government is a price-taker. What if, as in the
case of OPEC, it is a price-maker, or, in conjunction with others, can
through its own behavior influence the rate at which the asset will appre-
ciate in the ground? Then a high marginal utility for resource revenues which
in the "simple" case would have produced a rapid rate of exploitation might
induce a greater willingness to slow the rate of production (i.e., take
larger risks in the common producer interest) in an effort both to slow deple-
tion and to drive up present and future prices for the resource. On the
economics of exhaustible resources, see H. Hotelling, "The Economics of
Exhaustible Resources," Journal of Political Economy (April 1931); Robert M.
Solow, "The Economics of Resources or the Resources of Economics," The
American Economic Review vol. 64, no. 2 (May 1974); and Carlos Diaz-
Alejandro, "International Markets for Exhaustible Resources, Less Developed
Countries, and Transnational Corporations," draft, Yale University, November
1976.

investment offers the greatest economic return, however, it is clear that
social and political pressures for greater government expenditure will not
only continue at a high level but increase steadily in the future. Like
Saudi Arabia, Iran is discovering that it requires large amounts of spending
to maintain adequate public sector services (water, sewage, transportation,
power) in a country undergoing rapid social change.[27]   Moreover, there are
at least four areas that will demand larger public outlays in the future than
in the past simply to maintain the current level of political stability.
First, food subsidies: the Iranian government currently picks up the difference between controlled prices for foodstuffs at the retail and the producer
level. In 1974/75 this amounted to $1.2 billion (not foreseen in the Plan),
and has climbed since then. This has helped hold down inflation and cater
to the protests of consumers. But the price paid to farmers has not been
sufficient to bring forth the needed increase in output and has, besides,
created rural discontent. Iranian planners are convinced that producer
prices must rise, and that only a portion of the increase can be passed on to
consumers, thus swelling the current $1-$2 billion deficit. In addition, there
are specific nutrition programs, such as free lunches for school children,
whose costs will rise. Second, rural development: the boom in the urban sector
of the economy and the slow start, thus far, of effective rural service programs has meant a deterioration in many rural areas as doctors, teachers, owners
of equipment, and so forth have been bid away to the towns. This deterioration,
plus the discontent cited above from fixed prices for agricultural products

---

[27]Even in the provision of oil-fired electric power, for example, some
industries reported eighteen blackouts in the first three months of 1977,
several for as long as half a day! "Growing Pains: Despite Its Oil Money,
Iran's Economy Suffers From Many Shortages," Wall Street Journal, April 11,
1977.

with rising costs for fertilizer and other inputs, threaten to undermine the Shah's traditional bedrock of support among the peasantry. The situation can be moderated, or reversed, only with expensive programs of rural development. Third, <u>public housing and urban services</u>: the Fifth Plan left the provision of housing largely to the private sector, which has concentrated on high-cost apartments and villas. In addition, as mentioned above, approximately 70 percent of the "public housing" expenditures that have been made have in fact been military construction. As a consequence, there has been a severe shortage of middle class and lower class housing, with rents rising at a rate of 200-300 percent per year since 1973. At the same time, only a small proportion of industrial and government workers have access to adequate medical facilities. Many observers feel that the Sixth Plan will have to concentrate much more explicitly on the improvement of urban welfare than the Fifth Plan has in order to dissipate social tensions as Iran goes from the current boom to slower and more cyclical growth. Fourth, <u>military spending</u>: paradoxically, for many Iranians who see the military budget as an obvious target for potential cuts,[28] military spending may become a source of growing demands on the public purse. The primary reason is that military salaries must keep pace with the possibilities available in the civilian sector, especially in high-skill areas of technical and managerial expertise.[29] Thus, even if expenditures for military equipment were to drop off considerably (which is by no

---

[28]Discussions with middle class and upper class Iranians in Tehran in June 1976 revealed a good deal of resentment about military spending as the source of fiscal and balance-of-payments difficulties as well as a prime cause of inflation.

[29]For a similar view, see Robert Mantel and Geoffrey Kemp, "U.S. Military Sales to Iran," A Staff Report to the Subcommittee on Foreign Assistance of the Committee on Foreign Relations, United States Senate, July 1976.

means certain),[30] military pay will probably have to rise sharply and
steadily. Finally, the program of construction and expansion of bases which
Iran has just begun will require at least the current level of expenditures
even if stretched out over a long period of time into the future.

Thus, Iran faces increasing pressure from new budgetary needs of intense
importance to the stability of the existing regime at the same time it is
having to adjust to the fact that the Fifth and Sixth Plans will not be able
to propel the country into the ranks of the middle-level industrial powers
before its petroleum export potential begins to decline.

This suggests that the marginal utility curve for energy revenues for
Iran might look something like the following:

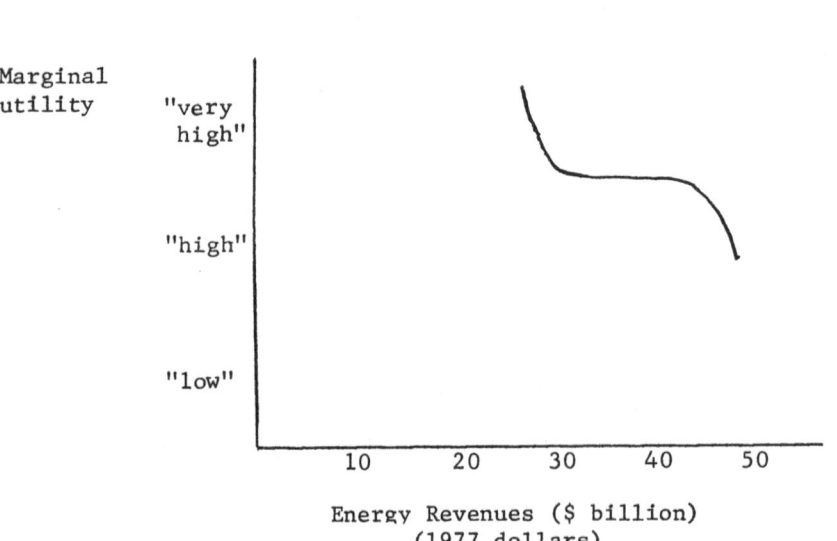

Iran

Marginal utility

"very high"

"high"

"low"

10   20   30   40   50

Energy Revenues ($ billion)
(1977 dollars)

---

[30]Cf. "Iranians Plan to Purchase $10 Billion in U.S. Arms," New York
Times, August 8, 1976. The purchase of 300 F-16 and 200 F-18 fighters, and
the creation of a "blue-water" navy to patrol the Indian Ocean, together with
the necessary logistical support facilities, would almost certainly preclude
a decline in expenditures on military equipment. With the rise in personnel
and construction costs, the natural result would be a steady upward slope for
the defense budget. Thus, even a military budget held constant at its current
high level would require Iranian defense chiefs, and the Shah, to delay or
give up plans for weapons systems that they clearly consider important for the
country's role as a regional power.

In short, any export level between now and the mid-1980s below full
capacity (minus domestic consumption) will constitute an acute "sacrifice"
for Iran unless oil prices rise substantially above current levels in
real terms.

3. Iraq

In contrast to other OPEC governments, Iraq never forecast that it would
accumulate large petrodollar surpluses. Rather, as Adnan Hamdani, Secretary
General of the Committee for Oil Affairs, pointed out on several occasions,
the oil production level would be allowed to vary according to what was
necessary to cover the government's revenue needs.[31] In 1972, prior to the
quadrupling of oil prices, INOC (the Iranian National Oil Company) had projected export capacity rising to at least 6.0 mbd by 1980.[32] After 1973,
Iraqi planners scaled down their forecasts to 4.0 mbd in response to the
higher per-barrel income "with a view to getting a better balance between
production and revenue needs."[33] They did not, however, cut back their oil
expansion programs in corresponding fashion. Rather, Iraqi strategy, as
related to the Petroleum Intelligence Weekly, was to maintain those programs
"unchanged at a high level, giving the country 'flexibility' and 'security'."[34]

In fact, petroleum sector expenditures have been increased to offset
the impact of inflation. Export capacity, according to the Oil and Gas

---

[31]Adnan Hamdani, Secretary General of the Follow-up Committee for Oil
Affairs and Implementation of Agreements, June 1975, quoted in Middle East
Economic Digest (MEED), December 26, 1975, p. 18.

[32]Middle East Economic Survey (MEES), June 20, 1975, p. 1.

[33]Adnan Hamdani, quoted in MEES, June 20, 1975, p. 4.

[34]Quoted in MEES, January 31, 1975, p. 3.

Journal could easily surpass the original target of 6 mbd by 1980.[35] At the same time that production capacity is being expanded, Iraqi estimates of the country's petroleum reserves, set at 35 billion barrels in the early 1970s, have been revised upward to "at least" 75 billion barrels, making Iraq second only to Saudi Arabia in the world in terms of oil reserves.

The first estimates of the expenditure targets to be included in the Third Five Year Plan before the 1973 oil price hike clustered around 5 billion Iraqi dinars (ID), or about $17 billion in constant 1973 prices.[36] Subsequently, the revised Third Five Year Plan, originally scheduled to begin in April 1975 has been publicized with dimensions double the size of the original version (ID 10 billion or $34 billion).[37] The revised Plan places much greater emphasis on industry and the building of infrastructure than did the original Plan, with agriculture (traditionally the focus of Iraqi development efforts) remaining almost unchanged. Outlays for industrial projects, including steel mills, chemical and petrochemical complexes, cement works, aluminum smelters, and manufacturing facilities, jumped about 300 percent (ID 1.4 billion to ID 4.0 billion). Revised infrastructure programs include, among other things, the quadrupling of electrical generating capacity, the construction of more than $1.3 billion in rail lines, and the improvement of urban facilities such as public transport and sewage.

Yet, in the interval between 1973 and 1977, international heavy construction costs have risen 45 percent to 85 percent, and local construction costs

---

[35]The Oil and Gas Journal, October 13, 1975.

[36]Hik-mat al-Azzawi, Minister of Economy, speaking at the opening of the Baghdad International Fair, October 1, 1974, as reported in MEES, October 11, 1974, p. 7.

[37]The final version of the Third Five Year Plan had not been published in official form as of the end of 1976. The analysis presented here is based on figures released to MEES, MEED, and the Economist Intelligence Unit, Quarterly Economic Review: Iraq, Annual Supplement, 1976, and Iraq, No. 2-1976.

within Iraq have increased at least 150 percent.[38]   (Since many prices in
Iraq are controlled by the government, the true escalation in construction
costs may be considerably higher.)   Thus, in the aggregate a Five Year Plan
of ID 10 billion begun in 1977 will probably buy less in terms of development
projects than the original Plan with programs amounting to ID 5 billion.

Indeed, in the extra two years it has taken to revise the third Plan
(it was supposed to follow on the Second Plan in April 1975) Iraqi construction costs have risen at least 91 percent.[39]   As a consequence, Iraqi planners
will get 48 percent less for their ID 10 billion development budget beginning
in 1977 than they would have projected two years earlier.   Thus, Iraq will
have to spend as much as its authorities originally estimated to get half of
what they anticipated.   Put differently, rising costs have already forced
the Iraqis, like the Saudis and the Iranians, to cut their expectations in
half.

At the same time, Iraq's past record of not spending all of the funds
approved for its development budget within the period of the official allocation will be much less likely to be repeated in the future.   In the past,
Iraqi budgetary performance was conditioned by two factors: first, the
country was largely cut off from the West and relied on an antiquated indigenous administrative structure to carry out public programs; second, the
country focused its Plan efforts on rural development.   The present emphasis
on large-scale, foreign-engineered and foreign-managed projects in industry
and infrastructure suggests that Iraq will have much less trouble spending
money rapidly and on time.

---

[38]Bulletin, Central Bank of Iraq, no. IV, 1975, pp. 41-42, and more
recent updates.

[39]Ibid.

If, in fact, Iraq contents itself with a plan whose total investment
does not exceed ID 10 billion (which is problematic),[40] while maintaining
its 1976 level of ordinary and state agency expenditures (which is highly
problematic), its annual budget will run at a level of approximately the
following:

Plan budget (annual)	ID 2.000 billion	$6.760 billion
State agencies (annual)	ID 2.075	7.014
Ordinary budget (annual)	ID 1.477	4.992
TOTAL	ID 5.552	$18.766

And, as in the case of Iran and Saudi Arabia, Iraqi expenditures will
decline in purchasing power each year by the amount the country's inflation
rate exceeds the OECD rate. The country's "inflation premium" for heavy
construction in 1976 was approximately 17 percent.

There will be strong pressures to increase expenditures in three parts of
the consolidated budget. First, Iraq faces the same problem as Iran with
regard to mounting food subsidies. The government in Baghdad is eager to
stimulate agricultural production, prevent rural-urban migration and maintain
support among the peasantry in the face of rising costs for farming inputs
(fertilizer, etc.). This requires high prices for food products. At the
same time, it maintains an artificial ceiling on food prices in the urban areas.
Consequently, the projection for food subsidies (including imports) must point
sharply upward.

Second, there is a growing need for public housing in Baghdad, Mosul,
and Basrah, exacerbated by a history of tight rent controls and a very
restricted private construction sector in the country's socialist economy.

Third, military expenditures have been running at a level of "only"

---

[40] In February 1976, Sabah Kachachi, director-general of the industrial
department in the Ministry of Planning, asserted: "We have been spending
at a rate of 2 to 2-1/2 billion dinars a year on development and we have no
intention of slowing down now. That means that there will be definitely more
than 10 billion dinars ($33 billion) in the new five-year plan." Mideast
Markets, February 16, 1976.

$1.740 billion (1975 annualized) in comparison to $6.4 billion for Saudi Arabia and at least $7.6 billion for Iran during the same period. If Iraq decides to re-equip its forces and modernize its bases with even a fraction of the intensity that its two traditional rivals have been exhibiting, one could predict a budget several times current outlays.

The following representation of Iraqi oil revenue needs assumes that Iraqi planners do not want to raise plan investment above ID 10 billion, that pressures for additional housing and food subsidies are ignored, that Iraq decides not to engage in a major foreign aid giving exercise, and that military expenditures rise to no more than $3 billion per year. The resulting annual budget (approximately $20 billion, of which $19.3 will be financed from petroleum revenues) would require an export level of 4.1 mbd in 1980 with constant 1977 oil prices.[41] This study will assume that Iraq in fact does not push its export capacity above 4.0 mbd by 1980 (unless oil prices should fall) but relies instead on funding its fiscal shortfall (half a billion dollars per year) through deficit financing.

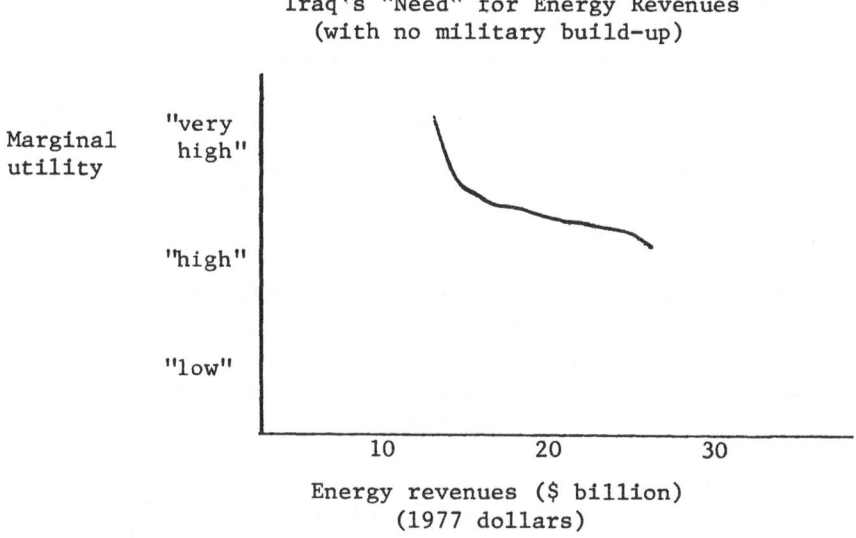

Iraq's "Need" for Energy Revenues
(with no military build-up)

[41]This assumes that the post-July 1977 price for Basrah crude is $13.32, for Kirkuk crude $13.59 with production costs of 20¢ per barrel for Basrah and 65¢ per barrel (including tolls) for Kirkuk.

4. Venezuela

Venezuela's plans for steady internal development (8.2 percent per year) to 1980, coupled with a substantial regional aid program, have been predicated upon a "conservationist" approach to petroleum exploitation. With a production capacity between 3 mbd and 4 mbd, output is scheduled for 2.2 mbd through the end of the decade, a level considered adequate to provide a major portion (52.7 percent) of the financing for the Fifth Plan.[42] This conservationist constraint, however, has resulted in a steady scaling down of the number of programs and projects.

Between the end of 1973 and the end of 1975, the purchasing power of revenues spent on major industrial projects (which make up more than 40 percent of the Plan) declined by approximately 57 percent, leading to the first major reduction of expectations.[43] Even so, the Plan, as originally circulated in Caracas, was not self-financing. Rather, it anticipated an increase in public indebtedness of $14 billion by 1980, or more than four times the level at the end of 1975 ($3.255 billion). This so shocked the Venezuelan business community when the planning minister, Gumersindo Rodriguez first offered the Plan in May 1975, that its publication was delayed (after President Carlos Andres Perez had already announced the goals to the nation), and almost $10 billion in programs were simply lopped off.[44] The resulting Plan is still not

---

[42]Of this, 2.0 mbd will go for exports, while 0.2 mbd will be consumed domestically.

[43]Index of construction materials December 1973 to December 1975, Central Bank of Venezuela. Domestic materials rose 71 percent. Imported materials rose 44 percent.

[44]Latin American Economic Report, April 9, 1976, pp. 59-60 and May 14, 1976, p. 75.

self-financing but requires a cumulative deficit of $4.3 billion.

Even the range of programs now contained in the Plan depend, however, upon some tenuous assumptions about the future:[45]

1) A massive increase in non-petroleum tax revenues. Nonpetroleum tax revenues jumped by 50 percent in 1975, due mainly to a dramatic rise in income tax and customs collections. The Plan is predicated upon another rise of nearly 50 percent (from 9.3 billion bolivares to 13.5 billion bolivares) between 1975 and 1980. Twenty-three percent of the Plan's financing ($13.4 billion) must come from these new tax collections, in order to limit the increase in public indebtedness to $4.3 billion.

2) Large net savings by Venezuela's state enterprises. Public sector enterprises are counted on to generate 39 billion bolivares ($9.2 billion) over the course of the Plan, at a rate by 1979 (10 billion bolivares per year) more than twice the contribution of 1976 (4.8 billion bolivares per year).

3) A 9 percent per year expansion in the public income generated by iron ore exports (via increases in price and volume and/or reductions in production costs).

4) The ability of the government to maintain current expenditures at a level no more than 87 percent of capital expenditures. The corresponding ratio for the period 1971-75 was 162 percent.

5) An increase in the cost of the social and economic programs contained in the Plan of no more than OECD rate of inflation.

The probability one assigns to successful financing of the Fifth Plan depends upon the confidence one has regarding the above assumptions. Should these assumptions prove to be overly optimistic, the Venezuelan

---

[45]Fifth National Development Plan, Government of Venezuela.

government can, of course, cut back its planned outlays beyond the
$10 billion in projects already dropped and show even more hesitation toward
continuing the regional lending programs than it demonstrated during 1976.

If oil prices do not rise in real terms, it is likely, however, that
at some point before the end of the decade Venezuela will begin searching
for new sources of revenue. There are several candidates: 1) the private
business sector in Venezuela; 2) personal incomes in Venezuela; 3) the large
tax exempt sector in Venezuela. Given the domestic reaction to encroachment
on any of these three, however, the more likely target is the petroleum
sector itself.

Venezuela's conservationist stance since 1973 has been justified domestically with arguments that 2.2 mbd per day would provide oil sufficient for
both domestic consumption and government revenues, that higher production
levels might not be technically feasible after the nationalization of the
foreign holdings in 1975, and that the country's oil reserves were perilously
nearing exhaustion. All three of these arguments will probably prove without
foundation by 1980. The technical competence of PETROVEN, bolstered by
service contracts with the former foreign owners, is expected, after an initial
shake-down period to be high. The Ministry of Mining and Hydrocarbons raised
its estimate of petroleum reserves in 1975 by 38 percent (13.8 billion barrels
to 19.0 billion barrels) due simply to the kinds of secondary recovery tech-
niques that have become feasible at the post-1973 price for oil. Finally,
with a concentrated exploration effort, there are prospects for new finds of
oil (not counting the heavy tar belt of the Orinoco).

For these reasons, it seems reasonable to predict that Venezuela will
have both the capacity to export at a level of at least 3 mbd (including
natural gas and products) in 1980 and a strong motivation to do so. Within
the analytical framework of this chapter, the marginal utility of revenues

at that level (3 mbd x 10.99 per barrel, the estimated government "take" for late 1977) will be high in 1980 and rise sharply for any revenue level less than that.

Venezuela's "Need" for Energy Revenues in 1980

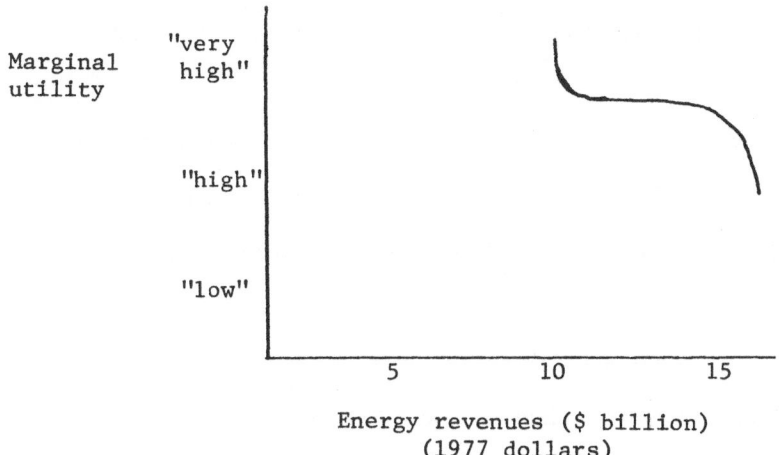

Energy revenues ($ billion)
(1977 dollars)

5. Kuwait, Libya, and the Small Gulf States

The budgetary needs of Kuwait and Libya, including foreign aid and grants to other Arab states, are somewhat obscure as a guide to their probable production (or price) preferences. In the three years following 1973, however, the two have repeatedly protested the decline in their exports and begun to discount their own oil so as to regain their market shares, when their sales fell below about 1.8 mbd.[46] Libya's sense of its own "needs" is particularly unpredictable. With a prior export record above 3 mbd, the director of the Libyan National Oil Company, Mr. Muntassar, has said that the government aims to increase oil production in the late 1970s to a level of about 2.4 mbd, a figure well within the estimates of the country's capacity

[46]Industry sources.

made by several international oil companies.[47] To be on the conservative side, however, this chapter will take 1.8 mbd (at 1976 prices) as representing the preferred export level below which both Libya and Kuwait will feel they are being called upon to make a sacrifice. At that level Libya will probably be running a fiscal and balance-of-payments deficit, and Kuwait will be producing the minimum amount of associated gas necessary to sustain oilfield use and run public services.

The revenue "needs" of Abu Dhabi are also opaque. Of earnings totaling approximately $4 billion in 1975, the country accumulated a surplus of only $500 million. Abu Dhabi finances almost the entire common budget of the United Arab Emirates ($2 billion in 1976), and spent nearly $1.6 billion in 1975 on internal programs ($800 million on "economic development," $500 million on defense, $300 million on current expenditures).[48] Foreign aid allocations of the government have run between $500 million and $1 billion since 1973. On the basis of these calculations, one official of the Abu Dhabi oil ministry has argued that the country would be reluctant to reduce exports below 1.5 mbd (with a capacity near 3.0 mbd by 1980) unless "it was taken in the context of OPEC and involved all member states making an equal sacrifice through export cuts."[49]

This study adopts the lower figure of 1.0 mbd as the export preference representing Abu Dhabi's perception of its own needs. Three arguments would seem to support this: first, the production of Abu Dhabi's traditional rivals in 1980, Dubai (0.4 mbd) and Qatar (0.7 mbd) will approximate this level;

---

[47]The New York Times, September 30, 1976.

[48]Ibid.

[49]Hareb Al-Damarki, "Abu Dhabi's Oil Pricing Policy," seminar paper, Johns Hopkins University School of Advanced International Studies, May 7, 1976, p. 10.

second, Abu Dhabi's influence in Emirate and broader Arab affairs will be
constricted at a lower production level; third, Abu Dhabi has appeared in the
past to discount its export prices shortly after production dropped below
the 1.0 mbd level.[50]

6. Nigeria, Indonesia, Algeria, Ecuador, and Gabon

This study assumes that the demands on the public purse in Nigeria,
Indonesia, Algeria, Ecuador, and Gabon are such that the countries will not
willingly want to carry any spare capacity (with constant real oil prices)
in 1980. Thus, their "export preferences" will be the level of production
that they can reasonably expect to sustain, less local consumption:

Export Capacity in 1980
(constant real oil prices)

	Million barrels/day oil equivalent
Nigeria	2.6
Indonesia	1.8
Algeria	1.4
Ecuador	0.3
Gabon	0.2

Note: These include natural gas exports as well
as exports of natural gas liquids, and products.
For Nigeria and Indonesia, these estimates are
lower than those of some other analysts.

7. OPEC Revenue Needs and Export Preferences

From the preceding estimates, it is possible to construct the following
table of revenue needs and export preferences for the OPEC countries (table 4).
The revenue "need" is the approximate level of earnings from energy exports
below which political elites must sacrifice programs that the preceding analysis suggests they consider "important" for domestic stability, economic development, social welfare, national defense, or international influence. This

---

[50]Industry sources.

Table 4

REVENUE NEEDS AND ENERGY EXPORT PREFERENCES IN 1980

	Revenue need ($ billion)	Export preference (mbd oil equiv.)	Energy export capacity (oil plus natural gas) (mbd oil equiv.)	Installed oil capacity (as of Jan. 1977) (PIW, Jan. 31, 1977)
Saudia Arabia	39.9	9.5	13.0	11.8
Iran	32.4*	6.9	6.9	6.7
Iraq	18.8*	4.0	5.0	3.1
Venezuela	11.0	3.0	3.5	2.6
Nigeria	9.9*	2.6	2.6	2.7
Indonesia	6.8*	1.8	1.8	1.8
Algeria	5.3*	1.4	1.4	1.1
Ecuador	1.1*	0.3	0.3	0.3
Gabon	0.8*	0.2	0.2	0.2
Kuwait	8.3	1.8	3.5	3.3
Libya	8.3	1.8	2.6	2.5
UAE	6.4	1.4	3.0	2.4
Qatar	3.2*	0.7	0.7	0.7
Total	152.2	35.4	44.5	39.2

Assumption: constant real 1977 prices.

*Note: The revenue "needs" of these countries are constrained by their probable export capacities in 1980. Should the price of oil rise in real terms, they all would still want to export as close to capacity as possible until energy export receipts reached a level substantially higher than the amounts indicated in the first column above.

chapter does not argue that they will refuse categorically to sacrifice such programs. On the contrary, the common interest that they all have in the success of OPEC suggests that they should be ready, rationally, to sacrifice many such programs if the survival of the producers' association is at stake. The aim here is to provide some point of reference at which pressures mount to shift a portion of the necessary production cutbacks onto other OPEC members, or to devise together a strategy to raise oil prices faster than the OECD rate of inflation. The revenue calculations for Saudi Arabia, Iran, Iraq, and Venezuela are based on government "takes" given in the text. For the other OPEC countries, industry estimates of producer government "takes," adjusted to late 1977 levels and weighted appropriately among types of crudes, are the following: Nigeria ($13.00), Indonesia ($12.60), Algeria ($12.40), Ecuador ($12.30), Gabon ($7.00), Kuwait ($12.00), Libya ($13.00), UAE ($12.30), Qatar ($12.60).

The contention of this chapter is that an aggregate export level below 35.4 mbd in 1980 (with constant real 1976 oil prices) will produce problems of increasingly severe difficulty for OPEC as the group tries to apportion the aggregate revenue squeeze among its members. But the reader should recall that the figure of 35.4 mbd is in fact a point on a continuum of mounting internal tension: it represents neither the moment of ultimate agony for the producers, nor the first sign of strain and dissatisfaction with export earnings. Even if all of the "export preferences" given above are met: Saudi Arabia will have a Plan still fifteen to twenty years from completion, mounting social welfare expenses, a sharp fall-off in the number of new contracts let with the aid of indigenous consultants, and declining financial reserves; Iran will have a fragile industrial base, a military complex well short of the Shah's desires, and potentially serious problems of domestic political tension; Iraq will be making steady (if unspectacular) progress in industry and agriculture but be locked in a position of decisive military inferiority vis-a-vis its two traditional rivals in the Gulf; and Venezuela, Nigeria, Indonesia, Algeria, Ecuador, and Gabon will all be seeking higher revenues for what their governments consider pressing social and economic programs.

Will OPEC have to contain itself to "only" 35.4 mbd in 1980? Will it be able to export more than that and never have to face the problem of rising internal tensions? Or will it have to compress its spending expectations far below the "revenue needs" indicated above (if prices do not rise)? These are the questions we turn to in the next chapter.

III. OPEC EXPORTS AND UNWANTED SPARE CAPACITY

The study has hypothesized that the farther OPEC revenues lag behind the "needs" of its members, especially the "needs" of members who have idle capacity, the greater the adjustment problem within OPEC and the greater the tension among the member governments. This does not mean that OPEC will break up spontaneously under the strain of adjusting market shares among its members. Indeed, as we shall see, the escalation of disputes about the distribution of a given level of revenues may constitute the catalyst that propels OPEC to perfect its structure as a cartel, prorationing output explicitly, with higher prices being the reward for agreement on who must accept idle capacity. But before jumping to possible OPEC market-sharing strategies, it is necessary first to estimate the plausible range of demand for OPEC exports. This chapter attempts three tasks. First, it constructs a "base case" of supply and demand for energy in the non-Communist world with the aim of deriving a residual demand for OPEC exports. Second, it allows both supply and demand to fluctuate within the extremes of plausible variation so as to indicate the likely range of demand for OPEC exports. (For each of these exercises oil prices are assumed to remain constant in 1976 terms to 1980.) Third, the price assumption is allowed to vary, from 25 percent lower than current levels to 50 percent higher by 1980.

The aim of this chapter is not to try to predict the exact OPEC export figure in 1980, but rather to show how more and less probable projections for energy supply and demand affect the prospects for tension within OPEC. This should provide a framework within which subsequent analysts can place later revised estimates of supply and demand.

On the demand side, the "base case" assumes that GNP in the non-Communist developed countries grows from 1975 to 1980 at an average of 5.0 percent per year, with an income elasticity of aggregate energy consumption of 1.0, a price elasticity of -0.15, and a lag effect as an energy-intensive capital stock is replaced by less energy-intensive capital goods that produce an aggregate 4.0 percent per year growth in energy. It should be noted that these growth rates for both GNP and energy consumption are higher than most current forecasts. (See table 5.)

Table 5

(NON-COMMUNIST) DEVELOPED COUNTRIES ENERGY DEMAND[1]
(million barrels per day oil equivalent)

	1972	1973	1974	1975 (est.)	1976 (est.)	1980 (est.)
OECD	67.2	70.9	70.0	68.7	71.5	83.6
Non-OECD countries	1.8	1.8	1.7	1.6	1.7	1.9
Bunkers not included in OECD statistics	0.5	0.5	0.5	0.6	0.6	0.7
Total	69.5	73.2	73.2	70.9	73.8	86.2

On the supply side, the "base case" takes a markedly more cautious view of the prospects for energy production in the developed countries than many other forecasts due to fears about nuclear power, concern about the environment, and uncertainty over long-term energy policy (especially in the United States).[2] (See table 6.)

---

[1] These data come from OECD, Statistics of Energy (Paris, 1976); United Nations, World Energy Supplies, Series J, No. 19 (New York, 1976); official sources for South Africa and The Commonwealth of Puerto Rico.

[2] The OECD, for example, projected in 1975 that its members would reach 21.8 mbd in 1980 for oil production at $9 per barrel (constant 1972 prices), 19.7 mbd for natural gas production, 17.6 mbd for coal production, and 6.5 mbd for nuclear production (oil equivalent).

Table 6

ENERGY SUPPLIES IN THE DEVELOPED COUNTRIES[a]
(million barrels per day oil equivalent)

		1973	1974	1975	1980
1.	Oil (including natural gas liquids)				
	United States	11.0	10.5	10.0	11.0
	Canada	2.1	2.0	1.8	1.6
	Western Europe	0.5	0.5	0.6	3.6
	Japan	0.0	0.0	0.0	0.0
	"rest"	0.6	0.6	0.5	0.6
	Total	14.2	13.6	12.9	16.8
2.	Coal				
	United States	7.1	7.3	7.9	8.5
	Canada	0.2	0.2	0.3	0.5
	Western Europe	4.6	4.3	4.5	4.3
	Japan	0.3	0.3	0.3	0.2
	"rest"	1.7	1.8	1.9	2.4
	Total	13.9	13.9	14.9	15.9
3.	Natural gas				
	United States	10.5	10.1	9.4	8.2
	Canada	1.3	1.3	1.3	1.4
	Western Europe	2.3	2.8	2.9	4.2
	Japan	0.0	0.0	0.1	0.1
	"rest"	0.1	0.1	0.1	0.1
	Total	14.2	14.3	13.8	14.0
4.	Nuclear (input basis)				
	United States	0.4	0.6	0.7	1.9
	Canada	0.1	0.1	0.1	0.2
	Western Europe	0.4	0.4	0.5	1.5
	Japan	0.0	0.1	0.1	0.5
	"rest"	0.0	0.0	0.0	0.0
	Total	0.9	1.2	1.4	4.1
5.	Hydro and geothermal (input basis)				
	United States	1.4	1.5	1.5	1.8
	Canada	1.0	1.0	1.0	1.3
	Western Europe	1.8	1.9	2.1	2.1
	Japan	0.4	0.4	0.4	0.4
	"rest"	0.1	0.1	0.1	0.2
	Total	4.7	4.9	5.1	5.8

[a]See table 7 for a comparison of these projections with those of the Organisation for Economic Co-operation and Development (OECD) the Congressional Research Service (CRS), the Central Intelligence Agency (CIA), Walter J. Levy, the Federal Energy Administration (FEA), and the International Bank for Reconstruction and Development (IBRD).

Oil

The oil forecast used here assumes that production in the "lower 48" of the United States will decline almost 10 percent (0.8 mbd) between 1975 and 1980 with Alaskan supplies, after some delay, reaching approximately 1.8 mbd by the latter date. Canadian production is also assumed to drop 10 percent during the same period. With regard to Europe, this analysis adopts the recent (conservative) estimates of Shell and BP for 1980 production from existing leases in the North Sea (2.1 mbd). British government figures are of course higher. It also assumes that Norway holds its production to 1.1 mbd, although Norwegian capacity should be more than 1.5 mbd. European onshore production edges downward to 0.4 mbd.

Coal

For the United States, this forecast includes 700 million short tons of coal from Appalachia, the Gulf states and the Midwest with an average heat content of 24 million Btu per short ton, and 100 million short tons of far "western" coal with an average heat content of 17 million Btu per short ton. This is considered compatible with strict strip mining regulations in both the East and the West as long as there is not an absolute prohibition on strip mining per se. For Western Europe this estimate includes a drop of more than 5 percent between 1975 and 1980 despite efforts in both Britain and Germany to reverse the decline.

Natural Gas

The estimate for the United States projects a 10 percent drop in U.S. natural gas production (1980 in comparison to 1975) even if interstate prices are decontrolled. This is larger in absolute terms than the decrease between 1970 and 1975 when prices were controlled.

Nuclear

The estimate of nuclear power capacity in 1980 has been constructed
to reflect concerns about safety not only in the United States but also in
Europe. It is lower, therefore, than 1976 World Bank calculations and fully
one-third below the 1975 OECD forecast.

Hydroelectric and Geothermal

In line with current construction plans the United States and Canada
are expected to increase hydroelectric and geothermal facilities by approximately 97.3 billion kwh and 56.0 billion kwh respectively between 1975 and
1980. Generating facilities elsewhere (including Western Europe and Japan)
are not projected to expand appreciably above 1975 levels.

To arrive at a residual demand for OPEC it is necessary to make two more
calculations: net non-OPEC LDC imports, and net communist energy exports.

Non-OPEC LDC Demand for Oil

This study follows the analysis of the World Bank on the energy needs
of the Third World, with the economies of the middle and higher income tier
growing at 5.5 percent per year (1974-80) and the lower income countries
growing at 4.5 percent per year. This produces an aggregate demand in 1980
of 11.2 mbd. (For details, see Energy and Petroleum in Non-OPEC Developing
Countries 1974-1980, World Bank Staff Working Paper No. 229, February, 1976).[3]
On the supply side, this study has reduced IBRD estimates for all energy
sources by 20 percent simply for the sake of caution. This produces an
aggregate supply of 8.4 mbd, with a gap of 2.8 which will have to be filled
almost entirely by oil imports. In addition to the across-the-board reduction of supply estimates, this analysis adopts low 1980 output estimates for

---

[3] World Bank, Energy and Petroleum in Non-OPEC Developing Countries
1974-1980 Staff Working Paper No. 229, February 1976.

Table 7

COMPARATIVE ESTIMATES of ENERGY SUPPLIES (1980) for NON-COMMUNIST DEVELOPED COUNTRIES
(mbd oil equivalent)

	OECD[a]	CRS[b]	CIA[c]	LEVY[d]	NEO[e]	IBRD[f]	MORAN
OIL							
U.S.	10.8	10.4	10.0	11.0	11.2	11.0	11.0
W. Europe	3.5		3.7	3.5	3.4	4.0	3.5
Canada	1.5		1.5	1.7	1.7	1.5	1.6
Japan	-		0.0	-	-	-	-
Other	NA		0.5	0.5	0.3(OECD Pacific)	0.5	0.6
TOTAL DEV'D	16.2(OECD)		15.7(OECD)	16.7	16.6	17.0	16.7
COAL							
U.S.	8.3	6.8	9.9	8.9	10.6	9.5	8.5
W. Europe	4.3		5.1	4.1	3.8	4.4	4.3
Canada	0.4		0.4		0.5	0.5	0.5
Japan	0.3		1.5		0.2	0.2	0.2
Other	NA				NA	2.4	2.4
TOTAL DEV'D	14.5(OECD)		16.9(OECD)	15.8(OECD)	15.1	17.0	15.9
NAT'L GAS							
U.S.	8.4	3.5	9.3	8.2		8.6	8.2
W. Europe	3.9		4.6	3.6		4.3	4.2
Canada	1.4		1.1	1.4		1.5	1.4
Japan	0.1					{0.2	0.1
Other	NA						0.2
TOTAL DEV'D	13.9(OECD)		15.4(OECD)[g]	15.6(OECD)		14.6	14.0
NUCLEAR							
U.S.	2.2	2.3	1.7			1.9	1.9
W. Europe	1.6		1.4			1.7	1.5
Canada	0.2		0.2			0.2	0.2
Japan	0.4		0.4			0.5	0.5
Other	NA					NA	0.0
TOTAL DEV'D	4.3(OECD)		3.6(OECD)[g]	4.1(OECD)		4.3	4.1
HYDRO AND GEO							
U.S	1.7	1.7	1.7				1.8
W. Europe	1.8		2.1				2.1
Canada	1.1		1.2				1.3
Japan	0.4		0.4				0.4
Other	NA						0.2
TOTAL DEV"D	5.3(OECD)		5.4(OECD)[g]	6.0(OECD)			5.8

[a]Organisation for Economic Co-operation and Development, World Energy Outlook (Paris, 1977).

[b]Congressional Research Service, Project Interdependence: U.S. and World Outlook Through 1990, by Dr. H. T. Franssen, for the Senate Committee on Energy and Natural Resources (Washington, June 1977).

[c]Central Intelligence Agency, The International Energy Situation: Outlook to 1985 (Washington, April 1977).

[d]Walter J. Levy, S.A., "Saudi Arabia's Approaching Choice," July 1976, and OPEC in the Medium-Term, May 1976.

[e]Federal Energy Administration, "National Energy Outlook," draft, January 1977.

[f]International Bank for Reconstruction and Development, "International Energy and Petroleum: Prospects to 1985," by J. Foster and R. Vedavalli, Commodity Paper No. 21 (May 1976).

[g]Excludes Australia and New Zealand.

Mexico (1.4 mbd) for Egypt (0.6 mbd, with many experts predicting 1.0 mbd), and for India (0.4 mbd). Thus this calculation assumes both high economic growth and relatively low energy output in the Third World. Many analysts have estimated Mexican oil production in 1980, for example, to lie in the 2.0 mbd range.[4]

Net Communist Exports

This analysis assumes that the Soviet bloc becomes a net importer of oil (0.3 mbd) in 1980, only slightly offset by net exports of natural gas running at 0.1 to 0.2 mbd oil equivalent and net exports of coal at 0.5 mbd oil equivalent. It assumes that Chinese oil exports reach no more than 0.5-0.6 mbd by 1980, although some observers have estimated as high as 1.0 mbd. These calculations reflect the public pessimism by the CIA about Soviet and Chinese oil performance.[5]

The resulting "base case" balance for OPEC is shown in table 8. Thus the OPEC governments can expect to derive revenues from approximately 30.4 mbd in energy exports (i.e., including natural gas and petroleum products) that they sell abroad.[6]

---

[4]Walter Levy, for example, estimates Mexican oil output in 1980 to be 1.8 mbd. (OPEC in the Medium-Term, Walter J. Levy S.A., May 1976, p. B-2.) The CIA puts 1980 Mexican production at 2.2 mbd (April 1977).

[5]Central Intelligence Agency, International Energy Siutation: Outlook to 1985 (Washington, April 1977). As recently as October 1976 the CIA was predicting exports of almost 1 mbd in oil alone to the West from the Soviet Union. "Outlook for Soviet Energy," Soviet Economy in a New Perspective, U.S. Congress, Joint Economic Committee, October 14, 1976.

[6]Domestic consumption in the OPEC countries will amount to nearly 2.2 mbd in 1980. The government revenues derived from domestic sales (at substantially lower than the OPEC price) have already been included in the calculation of nonenergy export tax receipts for the individual OPEC countries.

Table 8

BALANCE SUPPLY/DEMAND FOR ENERGY IN 1980
(million barrels per day oil equivalent)

I. Demand		
1. Developed countries (including bunkers)		86.2
2. Non-OPEC developing countries net demand		2.4
	Total	88.6
II. Supply		
1. Oil		16.8
2. Coal		15.9
3. Natural gas		14.0
4. Nuclear		4.2
5. Geothermal and hydro		5.8
6. Miscellaneous: re-refined oil, solar, synthetics		0.5
	Subtotal	57.2
7. Communist net exports (oil, coal, natural gas)		1.6
	Total	58.2
III. 1. OPEC energy exports (oil and natural gas) (I. minus II.)		
	Total	30.4

When matched with the "energy export preferences" calculated in the preceding chapter, this means that by 1980 OPEC will have to hold not only 14.1
mbd in spare oil and gas capacity but 5.0 mbd of spare capacity in the hands
of governments who "need" the revenues those idle facilities could bring.
This suggests a great deal of tension within OPEC over the question of how

to allocate internal market shares.[7]

1980 Unwanted Spare Capacity
(mbd oil equivalent)

Energy export preferences	35.4
Demand for OPEC exports	30.4
Unwanted spare capacity	5.0

The above calculation of 5.0 mbd held by countries for whom the marginal utility of extra revenues is high is based on the "base case" of supply and demand given earlier in this chapter. How far could the international energy equation diverge from that "base case" before the balancing problem would disappear, or at least recede to easily manageable dimensions?

The following cases represent alternative projections for supply and demand of varying degrees of plausibility.

Case I: Nuclear power is held to an output 20 percent below the (conservative) "base case" figure of 4.1 mbd oil equivalent in 1980, and the expansion of coal is able to take up only half of the slack. Result: demand for OPEC oil increases 0.4 mbd, and "problem spare capacity" decreases to 4.6 mbd.

Case II: The nuclear power estimate is reduced 20 percent as in Case I, but due to strip mining and environmental restrictions coal is not able to pick up any of the slack. Result: demand for OPEC oil increases 0.8 mbd, and "problem spare capacity" decreases to 4.1 mbd.

---

[7]The need to allocate spare capacity among countries that need the revenues means that continuing to build new facilities even when a proportion of them may operate at less than full throughput will be a high priority item for the OPEC governments. Indeed, as a "bargaining chip," new capacity may be one of the most cost-effective investments a government can make. Thus, Iraq will continue to build toward 5 mbd to legitimize its claim to the right to export 4 mbd while Iran and Saudi Arabia are being forced to cut back on their develoment programs. Only Saudi Arabia is estimated here to cut back on its construction of oil, natural gas, and petrochemical facilities in the late 1970s.

Case III: The conditions of Case II hold except that restrictions on coal expansion are so severe (primarily in the United States) that coal production itself falls 10 percent below the "base case" estimate.   (That would mean that U.S. coal production remains in 1980 at the same level--about 640 million short tons--as it was in 1975.)   Result: demand for OPEC oil increases 2.4 mbd, and "problem spare capacity" decreases to 2.6 mbd.

Case IV: Oil production lags drastically. By 1980 Alaskan output is less than 1.5 mbd and North Sea production is 20 percent below current Shell and BP estimates (2.0 mbd). U.S. production in the "lower 48" is 10 percent below 1975 levels (9.0 mbd). Result: demand for OPEC oil increases 0.7 mbd and "problem spare capacity" decreases to 4.3 mbd.

Case V: Energy output is a disaster. The worst of cases III and IV hit nuclear, coal, and oil production simultaneously, plus natural gas output lags 10 percent behind the conservative "base case" estimate. Result: demand for OPEC oil increases 3.1 mbd, and "problem spare capacity" decreases to 1.9 mbd.

Case VI: On the demand side, consumption of energy by the developed countries rises from the 4.0 percent per year growth rate hypothesized in the "base case" to 5.0 percent per year. Result: demand for OPEC oil increases 3.4 mbd, and "problem spare capacity" decreases to 1.6 mbd.

In summary, for one to conclude that OPEC will continue to enjoy a future with few internal strains from adjusting market shares among its members, one must expect some combination of the worst cases (V and VI) on both the supply and the demand sides simultaneously, i.e., a series of "disasters" in output plus a rate of growth in energy demand approaching 5 percent per year.

On the other hand, what if supply is more abundant than the "base case" or demand is more restrained?

Case VII: Oil production is moderately higher than the "base case" assumes.
Alaskan output reaches 2.0 mbd by 1980; North Sea production (UK and Norwegian
sectors) reaches 3.7 mbd; Mexico produces 1.6 mbd; Egypt produces 0.7 mbd.
Results: demand for OPEC oil drops by 1.3 mbd, and "problem spare capacity"
increases to 6.3 mbd.

Case VIII: Coal production is higher than the "base case" assumption.
In the U.S., output of "eastern" coal expands by 100 million short tons over
1975 production, instead of the 60 million short tons hypothesized in the
"base case" as a result primarily of operating some mines now on-line six
rather than five days per week. (At present, coal mining capacity is defined
on the basis of 235 operating days per year.) Australia, New Zealand,
South Africa, and Indonesia increase their production by 85 million metric
tons instead of the 35 million metric tons hypothesized in the base case.
China expands its output by 10 percent for export (47 million metric tons).
Results: demand for OPEC oil drops by 1.7 mbd, and "problem spare capacity"
increases to 6.7 mbd.

Case IX: Growth in energy demand is 3.5 percent rather than the 4 percent
per year hypothesized for the developed countries. Results: demand for
OPEC oil decreases 2.8 mbd and "problem spare capacity" increases to 7.8 mbd.

Thus, easily plausible variations of either supply or demand can greatly
expand the number of idle facilities that OPEC governments must carry in order
to balance supply and demand in 1980 at constant prices. As we shall note
later in the section on policy implications, this suggests that efforts to
stimulate domestic energy supplies or conserve domestic energy demand on the
part of the developed countries (especially the United States) will have a
dramatic impact in intensifying the market-sharing problem for the OPEC members.

Thus far all calculations have assumed that OPEC holds prices constant

in real terms (i.e., allows them to climb no more rapidly through 1980 than
the OECD rate of inflation). What if instead OPEC allows oil prices to drift
lower over the same period?

Case X: The price of oil declines 25 percent between the beginning of
1977 and the end of 1980. Results: paradoxically, a decline in petroleum
prices such as the one hypothesized here will probably exacerbate the spare
capacity problem for OPEC. Aggregate energy demand might rise (at the most)
2.6 mbd and non-OPEC energy supplies might fall 2.0 mbd, leading to a net
increase in OPEC exports of 4.6 mbd. But to meet a given expenditure target,
a member state would have to export higher volumes--i.e., an Iraqi or a
Venezuelan government willing to settle for a given rate of spending on its
development and defense needs with a certain export level at constant prices
would push (or want to push) for a higher export target if prices were to
decline by 25 percent. Among the OPEC countries that could expand production,
the exports needed to offset the drop in prices would be larger than the rise in
in demand for OPEC oil. Unwanted spare capacity grows from 5.0 mbd to
7.2 mbd.

To arrive at this result, I have assumed that with a steady price
decline the ratio of growth in energy consumption to growth in GNP among the
developed countries rises to its historical figure (about 1.0) by 1980.
(This is clearly a generous assumption: it means that the ratio of energy
growth/income growth would be the same in 1980 as if nothing had happened to
prices since 1972, whereas in fact prices will still be more than three times
higher than the 1972 level. Moreover, many analysts expect that evidence of
a long-term decline in the energy growth/income growth ratio will just be
appearing clearly in the late 1970s as the capital stock of the developed
countries is turned over.)

With a GNP growth rate of 5 percent for the developed countries to 1980,

energy demand will rise 2.6 mbd above the original "base case" estimate.
On the supply side, hydro, geothermal, nuclear, and natural gas forecasts
will change very little by 1980. On the basis of private oil company
projections, I have hypothesized that oil production might be 1.3 mbd lower
than the "base case" estimate (15.5 mbd) and coal production 0.7 mbd lower
(15.2), for a total decline of 2.0 mbd in supplies. This results in an
increase of demand for OPEC exports of 4.6 mbd (to 35.0 mbd).

To meet the revenue needs hypothesized in the "base case" at the new lower
price, the countries that have the capacity to expand production would have
to increase exports by 6.8 mbd, more than offsetting the 4.6 mbd increase in
demand. Saudi Arabia would have to export 12.7 mbd instead of 9.5 mbd, Libya
2.4 mbd instead of 1.8 mbd, Kuwait 2.4 mbd instead of 1.8 mbd, the UAE 1.9 mbd
instead of 1.0 mbd. Both Iraq and Venezuela would be exporting at their 1980
capacity. In toto, 2.2 mbd (6.8 mbd-4.6 mbd) must be suppressed to balance
supply and demand with the 25 percent price cut. When added to the "base
case" estimate, 6.4 mbd of spare capacity will be held by governments for
whom the marginal utility of the revenues foregone is high.

If the price decline were to continue, of course, the spare capacity
problem would work itself out, as one country after another raised its production to full output in an attempt to make up in volume what it was losing in
price. Such an evolution might leave the countries that have huge oil reserves no worse off in the present than they would be with higher prices
and lower levels of capacity utilization.[8] But it would skew the distribution
of benefits among OPEC members in such a way (notably toward Saudi Arabia,
Iraq, and the United Arab Emirates and away from Iran) that would be hard

---

[8]This sets at zero the discounted present value of the oil sales lost in
the future under the second alternative. In other words, it assumes
(unrealistically) that the countries involved place no value whatsoever on
conservation.

to accept. Moreover, the gains from capturing a larger share of the world energy market with lower prices would not be large enough--even for the individual countries receiving the greatest share of revenues--to compensate all the rest for their losses after their own expenditure targets were met. And political sensitivities would inhibit both the giving and the receiving of hypothetical side-payments among the major producers (e.g., from Saudi Arabia and Iraq to Iran) to maintain harmony within OPEC.

What about the opposite scenario--if OPEC raises oil prices through 1980? The problem of allocating market shares so that all countries can meet their revenue targets disappears only very slowly and never goes away entirely, as long as major recessions can be avoided.

Case XI: OPEC raises real oil prices steadily over the three year period 1978 through 1980 by 25 percent. This case hypothesizes a price elasticity of demand for energy of -0.1, an income elasticity of 1.0, a loss of income of 0.1 percent each year due to the price hike, and a price elasticity of supply over this period of 0.1. Energy demand drops 2.5 percent from the "base case" due to the price increase, and drops another 0.3 percent due to the decline in aggregate economic growth among consuming countries. This reduces energy consumption by 2.5 mbd in 1980. Energy supplies are hypothesized to increase by 0.6 mbd. The decline in demand for OPEC oil is 3.1 mbd, for a total of 27.3 mbd.

With the new 1980 price of oil ($16.65 per barrel in 1977 dollars), OPEC export preferences are:

	mbd	Revenues ($ billion)
Saudi Arabia	7.6	$39.9
Iran	6.9	40.5
Iraq	3.2	18.8
Venezuela	2.4	11.0
Libya	1.4	8.5
Kuwait	1.4	7.9
UAE	1.1	6.3

	mbd	Revenues ($ billion)
Qatar	0.7	$3.2
Nigeria	2.6	15.4
Indonesia	1.8	10.3
Algeria	1.4	7.9
Gabon	0.2	0.6
Ecuador	0.3	1.7
Total	31.0	

Note: The revenue needs of eight of the OPEC countries (Iran, Iraq, Nigeria, Indonesia, Algeria, Ecuador, Gabon and Qatar) are assumed to be sufficiently high that they would prefer to export at capacity if they could. Government revenues are calculated from the new price of oil minus local production costs.

Result: unwanted spare capacity declines from the "base case" level of 5.0 mbd to 3.7 mbd (31.0 mbd-27.3 mbd). This means that the frustration felt by several, or all, of the OPEC governments as they move slightly down their revenue marginal utility curve decreases--that is, their individual financial squeezes ease somewhat in comparison to the "base case." But they still must shut in 3.7 mbd that they would like to produce to satisfy their perceived revenue needs, and their preoccupation with having their colleagues rather than themselves do most of the shutting-in remains intense.

Case XII: OPEC raises real oil prices dramatically over the three year period 1977 through 1980 by 50 percent. This case hypothesizes a price elasticity of demand for total energy of -0.1, an income elasticity of 1.0, a loss of income each year due to the price hike of 0.2 percent, and a price elasticity of supply over this period of 0.1. Energy demand drops 5.0 percent from the "base case" due to the price increase, and drops another 0.8 percent due to the decline in economic activity. This reduces energy consumption by 5.1 mbd in comparison to the "base case." Energy supplies are hypothesized to increase by 0.6 mbd. The decline in demand for OPEC oil is 5.7 mbd, for a total of 23.4 mbd.

With the new 1980 price of oil ($20.81 per barrel in 1977 prices), OPEC export preferences are:

	mbd	Revenues ($ billion)
Saudi Arabia	5.4	$39.9
Iran	5.7	42.0
Iraq	3.2	24.0

	mbd	Revenues ($ billion)
Venezuela	2.0	$12.0
Libya	1.2	8.5
Kuwait	1.2	7.9
UAE	0.9	6.3
Qatar	0.5	3.4
Nigeria	2.6	18.5
Indonesia	1.8	12.4
Algeria	1.4	9.5
Gabon	0.2	0.8
Ecuador	0.3	2.0

Total 26.4

Result: unwanted spare capacity declines from the "base case" level of 5.0 mbd to 3.0 mbd (26.4 mbd-23.4 mbd). The revenue needs of Nigeria, Indonesia, Algeria, Ecuador, Gabon, and Qatar are assumed to be sufficiently high that they would prefer to export at capacity if they could. Government revenues are calculated from the new price of oil minus local production costs.

Variations on Case XII: One should note that an income loss of 0.3 percent per year (instead of the 0.2 percent per year hypothesized in Case XII) increases unwanted spare capacity to only 2.9 mbd, and an income loss of 0.4 percent per year increases unwanted spare capacity to only 3.0 mbd, and an income loss of 0.5 percent per year increases unwanted spare capacity to 3.1 mbd.

Thus, as long as the developed countries can use macroeconomic policies (fiscal and monetary) to keep their annual income loss moderate and as long as the medium term price elasticities of demand or supply do not change radically from the values hypothesized in Case XII (-0.1 and 0.1 respectively), OPEC countries can raise their prices dramatically before their actions become counter-productive.

Reasons why the estimates of spare capacity in cases XI and XII may be too low:

1) A price rise for energy of 25-50 percent by 1980 will doubtless have an inflationary impact. Since these estimates are given in constant prices, such inflation should not influence the calculations. Nevertheless, if, as

in the 1974/75 inflationary period, the prices of capital equipment, heavy construction, and food imports rise faster than the general world-wide price index, the budgets of the OPEC governments and hence their "export preferences" may be higher than estimated here.

2) Rising oil prices mean that the incentive to push the development of marginal capacity in the OPEC countries heightens. Thus, although this case assumes the same export capacity in OPEC as the "base case," this is probably an underestimation.

Reason why the estimates of spare capacity in cases XI and XII may be too high:

1) The assumption of chpater II that the high population "fringe" of OPEC (Nigeria, Indonesia, Algeria, Gabon, and Ecuador) will feel the need to produce at or near capacity may not be true with a 25-50 percent price rise in real terms. At some point as their revenues increase in that price range they may become relatively indifferent about additional earnings.

This chapter has hypothesized that an oil price rise of 25-50 percent over the three year period 1977-80 will not noticeably alter their preference to earn as much from their energy sector as possible. That may not be true.

OPEC to 1985

Projections to 1985 are subject to a much larger margin of uncertainty than forecasts to 1980. But the most plausible estimates possible with today's data suggests that with prices rising no more than 15 percent in 1977 and remaining constant in real terms thereafter the tension about how to distribute market shares within OPEC will extend well in the middle-1980s.

To calculate the demand for OPEC exports, this study reinterprets the "base case" forecast for 1985 made by the OECD in such a way as to add almost three million barrels per day. (Demand for energy is assumed to grow at 4 percent per year. See table 9.)

Table 9

1985 ENERGY BALANCES
(million barrels per day oil equivalent)

	OECD 1977	Base case for this study
I. OECD Demand	101.9	101.7
II. Non-OPEC Developing Countries Net Demand	-0.8	0.7
III. Other Countries and Miscellaneous Net Demand	1.9	1.9
Total	103.0	104.3
IV. OECD Supply		
1. Oil	17.7	17.7
2. Coal	17.3	17.9
3. Natural gas	15.5	15.5
4. Nuclear	9.3	7.3
5. Hydro, geo, other	6.1	6.1
Subtotal	65.9	64.5
V. Communist Net Exports (Oil, Coal Natural Gas)	2.0	0.0
Total	67.9	64.5
VI. Demand for OPEC Exports	35.1	39.8

There are three areas in which the "base case" taken for this study are more conservative than the OECD estimates: 1) nuclear power in the OECD states; 2) net energy imports for the non-OPEC developing countries; and 3) communist net exports. For nuclear power, this study adopts the methodology of the Standing Group on Long Term Co-operation of the International Energy Agency and assumes that nuclear plants which have not been approved, cited, and financed by 1977 will not be in operation by 1985. This reduces the contribution of nuclear power from 9.3 mbd in 1985 to 7.3 mbd, only a small portion of which can be made up by increased coal consumption, including (as the OECD report suggests) increased coal imports. For the non-OPEC LDC energy balances, this study assumes a 5 percent average annual rate of growth in energy demand from 1974 to 1985, for a total demand of 14.5 mbd in 1985. Aggregate supplies are expected to equal 13.8 mbd, nearly 10 percent below the OECD/IBRD projections of 15.0 mbd with Mexican production at 2.6 mbd (1.4 mbd exports), and Egyptian production at 1.3 mbd (0.8 mbd exports). If indigenous energy supplies do not reach the levels projected here, balance-of-payments constraints are expected to limit LDC economic growth rates rather than permit an increase in oil imports. Finally, with regard to Communist net exports, this analysis reflects the pessimism of the CIA about the prospects for the Soviet oil program but doubts that the Soviet bloc will have the hard currency to pay for the 3.5 mbd imports (cost $17.2 billion per year in 1977 dollars) that the CIA projects.[9] In 1976 the Soviet Union earned approximately $4.9 billion in energy exports (40 percent of its total hard currency earnings). This study limits the swing in the Soviet balance-of-payments to about $7 billion per

---

[9]Central Intelligence Agency, International Energy Situation: Outlook to 1985.

year by 1985, with the Soviets importing roughly the same amount (0.5 mbd) as the Chinese export. The rest of the Soviet short-fall in oil production, if any, will be reflected in greater conservation, switching to coal or gas, or ultimately slower domestic growth.

With regard to the export preferences of the individual OPEC members only Saudi Arabia and Iraq will have both the desire and the capacity to expand production significantly. This study hypothesizes that the growth in Saudi spending will peak around 1980 and taper off through 1985 to a rate of 8 percent per year. I have assumed that all of that will represent real increase, although the reader should note that Aramco (in contrast) forecasts its own composite costs for business operations on the peninsula rising at a rate of 15 percent per year in the early 1980s. If the latter turns out to be true, Saudi Arabia will be at a point of approximate "steady state" in budgetary expenditures from 1980 through 1985.

For Iraq, this study also postulates an 8 percent per year increase in budgetary outlays which, again, may mean no real growth in public spending. Since the analysis of Iraqi expenditures to 1980 did not include a major renovation of the country's armed forces to match the Iranian and Saudi build-ups (and by implication there could be no such major renovation by 1985 with a steady state budget), this hypothesis of a push toward 5.9 mbd in exports by 1985 should probably be considered quite conservative.

For the rest of OPEC, most of the other members are postulated as doing no better than expanding production to meet their own domestic consumption. For Iran that means making more intensive use of natural gas in the internal industrialization program. Only Libya, the UAE, and Qatar are seen wanting to expand exports (slightly) to earn more foreign exchange for purposes of exercising influence in international affairs.

1985 Energy Export Preferences
(mbd oil equivalent for oil, natural gas, and products)

Saudi Arabia	14.0
Iran	6.9
Iraq	5.9
Venezuela	3.0
Nigeria	2.6
Indonesia	1.9
Algeria	1.4
Ecuador	0.3
Gabon	0.3
Kuwait	1.8
Libya	2.4
UAE	1.8
Qatar	0.8
Total	43.1

1985 Unwanted Spare Capacity in OPEC
(mbd oil equivalent)

Energy export preferences	43.1
Demand for OPEC exports	39.8
Unwanted spare capacity	3.3

Thus, the necessity for governments who "need" additional revenues to hold a sizable amount of spare capacity idle does not disappear in the mid-1980s.

IV. IMPLICATIONS AND CONCLUSIONS

1. The expenditure projections set forth in section II indicate that the "absorptive capacity" of the OPEC countries, both individually and collectively, has grown in a very short span of time to be very much greater than early studies speculated. Moreover, if the projections included here err, it is probably on the side of underestimation.

2. These expenditure projections, when matched against the range of probable demands for OPEC exports in section III, undermine the belief that the low-population "core" of OPEC, in particular Saudi Arabia, can act as residual balancer for the producers' association. By 1980 the Saudis will not be able to cut back production "as much as necessary" to make supply and demand match. Internal needs, plus international "commitments" and "responsibilities," will push sectors of the ruling Saudi elite concerned with domestic stability, national defense or external influence to insist that other members of OPEC "allow" them a share of the market large enough to generate more than $35 billion per year revenues.[1]

3. Consequently, the "easy days" for OPEC as a semi-cartel with negligible problems of proportioning output will soon come to an end. With constant real oil prices they will probably begin to disappear in 1978 as the OPEC governments will have to decide who will absorb growing amounts of spare capacity. OPEC will have to face the problem of how to organize itself as a real cartel, assigning market shares explicitly, monitoring production, and disciplining cheaters.

---

[1] Nevertheless, Saudi financial reserves should remain high enough through the first half of the 1980s, even if oil prices rise no higher than the OECD rate of inflation, to allow the country to cut back production dramatically for a short period of time (three to six months) should political conditions dictate such an extreme measure.

4. The easiest way for the OPEC governments to solve the problem of obtaining agreement on market shares in the face of near universal revenue squeezes on the individual members will be to promise regular and dependable price increases as the reward for adherence to whatever prorationing scheme is adopted.

The primary constraint on the rate of increase in the price of OPEC oil will be the ability of the macroeconomic policies of the major oil-importing governments to accommodate such increase without falling into or aggravating recession. The more vulnerable the major consuming countries are to an oil price-provoked recession, the more cautious OPEC pricing policies will have to be. The more resilient the major consuming countries become, the more aggressively OPEC can push prices without its actions being counterproductive.

5. Paradoxically, however, OPEC, in the late 1970s will be heading in two directions at once: trying to strengthen and perfect its structure as a cartel while becoming more vulnerable to pressures of fragmentation. The intensification of the spare-capacity-cum-financial-squeeze predicament will raise the level of tension within OPEC as the members try to push the burden of production cutbacks, and hence budget cutbacks, off onto their neighbors and rivals. This setting will magnify the suspicion of each member that other governments, under the pressure of domestic politics, will try to fortify sagging exports by price shading. Given the extensiveness of the idle producing facilities that will exist across the range of prices hypothesized in section III, any exporter who does not respond rapidly in kind when others start to sell at a discount could find his market share eroded substantially. This will intensify the anxiety of every individual exporter, and complicate the efforts of the group to monitor and control cartel behavior rapidly and effectively.

6. The direction in which OPEC evolves--finding in adversity the catalyst for cohesion or the seeds of disintegration--will be profoundly affected by the energy policy that the oil-importing countries, preeminently the United States, adopt between now and 1980. Unlike the first three years of the energy crisis (1973-76) when a reduction of U.S. demand by several million barrels per day, as Secretary Kissinger initially pressed for, could be offset by the contraction of OPEC (e.g., Saudi) supply without so much as a financial murmur, efforts to restrain consumption and stimulate output outside of OPEC will in the future greatly complicate OPEC's difficulty in balancing supply and demand. This will hold true over a broad range of oil prices, from -25 percent to +50 percent of the 1977 price.

7. This analysis undermines the notion that the United States can continue to base its policy toward OPEC on the gratuitious goodwill of the Saudis to prevent further rises in the real price of oil (even in a weak market). The projection of revenue needs presented here suggests that the intra-OPEC struggles over price levels in 1977, including warnings by the Shah that expanded production by Saudi Arabia and the UAE constituted "an act of aggression" against Iran, may simply be the prelude to rising tensions among the OPEC members lasting into the 1980s. In such circumstances, the combination of internal financial needs and external pressure from fellow OPEC members may push Saudi policy--in the absense of countervailing pressures from the oil-importing states, especially the United States--in the direction of price hawks inside and outside the country. In a weak oil market, this will require prorationing, probably rationalized in the name of conservation.

8. If the United States wants to continue to make relations with Saudi Arabia the keystone of its hope for price moderation, it faces a two-fold challenge: first, to structure the multiple strands of the relationship with Saudi Arabia so as to ensure that the Saudis elect to meet their growing

revenue needs by expanding exports at constant prices rather than constricting exports at higher prices; second, to help insulate Saudi Arabia from the pressures and threats of the other OPEC members. Conditions are not unfavorable, in fact, for moving in the direction of an explicit bilateral agreement on prices and production that would not require either the United States or Saudi Arabia to do much more than it would like to do in its own self-interest anyway. First, in the wake of the Abqayq fire in early 1977, the Saudi elite's perception of its own vulnerability is acute. Many Saudi groups are convinced that the explosion was sabotage and they realize that they have neither the internal nor the external intelligence capability to protect themselves from future actions. They will have to rely upon U.S. government agencies for information and U.S. government help in designing an adequate security system to protect the nation's oil facilities. Second, the United States has enunciated an arms sales policy that limits the sale of massive amounts of weapons, especially sophisticated weapons, to "exceptional" cases. A schedule for delivery of Saudi military requests over the next eight years could be part of an agreement on oil price and production policy during the same period. Third, as this study has indicated, the exposure of Saudi Arabia to intense pressure on the oil price issue will increase over the medium term. The Saudis will not be able to avoid the hostility of either their fellow producers or their oil consumers if they want to proceed with their own expenditure plans. Reassurance that if they follow U.S. advice on price moderation they will receive U.S. support in the case of threats from Gulf neighbors could help them choose to side with the United States. In short, the United States might try to use those aspects of its relationship with the Saudis that the Saudis could least easily replace--intelligence-sharing, sales and training, and security support--to produce a medium term agreement on price and production policy.[2]

Whether either the U.S. government or the Saudi government could credibly commit itself to such an arrangement is highly uncertain. What is clear is that the United States cannot continue to rely on the spontaneous moderation of the Saudis to ensure that oil prices do not again begin to rise in real terms. Nor can it continue cavalierly to ignore, or finesse, security threats among OPEC members in the Gulf.

9. To the extent that one is dubious about the ability of the U.S. government to design, implement, and police an explicit bilateral agreement with the Saudis, however, one may want to explore other methods to set a ceiling on the cartel's power to dictate prices, to siphon off revenues from OPEC to the U.S. Treasury, and ultimately (perhaps) to undermine OPEC's ability to collude.

The analysis presented in this study suggests that the setting may be propitious for two such proposals: an import tariff (as recommended by Hendrik Houthakker), and an import quota action (as recommended by M. A. Adelman).

---

[2]The Saudis will clearly not want to give up their ability to impose an embargo and/or production cutback on the West (U.S.) in the event of another Mideast war. This will be a problem but not necessarily an insurmountable one. A "force majeur" clause could be included in the arrangement, defined to mean that the agreement might be suspended in the event of a Mideast war. This would mean that the U.S. defense supply and intelligence-sharing relationship would be suspended automatically as soon as the Saudis fiddled with oil production. This would give them pause, especially since, with an embargo or cutback, many observers would probably be urging a U.S. invasion anyway. The automaticity of the suspension of defense and intelligence arrangements would not require the president or the secretary to rattle his saber, or mutter publicly about strangulation to remind the Saudis of the seriousness of their action.

An import tariff. Hendrik Houthakker,[3] assumes that OPEC will push
prices in the direction of the price-quantity combination that maximizes
cartel revenues, subject to a minimum output constraint needed to ensure
OPEC cohesion without side-payments among the members. He argues that,
with the OPEC countries near their optimum point, an import tariff would
result in a pure transfer from exporting governments to importing governments.
OPEC could do nothing (e.g., raising their prices to offset the aggregate
revenue loss resulting from lower exports) without reducing earnings even
further and magnifying the strains among the members.

If, on the other hand, OPEC can raise real prices higher than the
current level before the effort becomes counterproductive, the United States,
argues Houthakker, should launch a preemptive strike by raising prices via
the tariff first. This will reduce energy consumption, stimulate domestic
production, and prevent the shift of an equivalent amount of resources to
OPEC. If energy prices are going up anyway, better us than them; we will
save the drain on U.S. resources (by refunding the earnings from our tariff
to our own populace), on the U.S. balance-of-payments, on U.S. economic
activity. Besides reducing the country's long-term economic vulnerability
to OPEC, the Houthakker proposal will also reduce the short-term import
vulnerability: the higher domestic prices produced by the tariff both
encourages conservation and stimulates output.

The import tariff proposal has the added attraction, according to
Houthakker, of aggravating the strains within OPEC as members with balance-
of-payments problems are tempted to discount from the present price to

---

[3]Hendrick S. Houthakker, The World Price of Oil: A Medium-Term Analysis
(Washington, D.C., American Enterprise Institute, 1976).

expand their market share. Houthakker hypothesizes, without detailed analysis, that 20 mbd is the lowest daily output from the Middle East and African producers consistent with maintaining OPEC cohesion in 1980. This output could be achieved in a way that maximizes OPEC revenues using what Houthakker judges are "pessimistic elasticities," by raising real oil prices one-third between 1976 and 1980. He argues that the IEA countries, led by the United States, should cut off this possibility for OPEC by imposing a tariff equal to such a price rise ($3.25 per barrel in 1976 prices) immediately.

U.S. import needs are large enough, according to Houthakker's calculations, to exercise a significant effect on OPEC output by varying an import tax unilaterally. The effectiveness of the proposal would be much greater, however, if the IEA countries could levy a common tariff.

The analysis presented in this study gives strong support to the Houthakker approach. By 1980, the Middle East and African members of OPEC will have revenue needs (and hence export preferences at current prices) equal to 30.3 mbd. Using Houthakker's pessimistic elasticities, these ten countries will have to absorb 10.3 mbd in unwanted spare capacity without cheating (one-third of the total they would like to export) in order to maintain the cartel intact. Alternatively, one can adopt elasticities much more pessimistic than Houthakker and still conclude that a $3.25 import tariff would put the cartel's cohesion under severe strain.

An import quota auction. In contrast to the import tariff proposal, an import quota ticket auction, as advocated by M. A. Adelman,[4] would aim at siphoning off revenues from OPEC and undermining the cohesion of the cartel without raising energy prices in the United States.

---

[4]M. A. Adelman, "Oil Import Quota Auctions," Challenge (January–February 1976).

Adelman proposes that the United States auction off oil import entitle-
ment tickets by sealed bid to cover the requirements each month. (The
government would not use the auction system to reduce either oil consumption
or imports.) Rather the U.S. would adjust the size of its offering continuously and sell tickets with variable maturities to ensure that there were
enough tickets in circulation at any point in time to meet the needs of
domestic refiners. The bidding would be open to anybody, with only the
deposit of a certified check required for qualification. The U.S. government
would keep the names of the bidders secret, and in any case encourage the
use of front companies and an active secondary market for the tickets. This
would mean that any exporter who wanted to keep his U.S. sales would have to
bid directly or indirectly, on the tickets. It would also offer any exporter
who wanted to expand his U.S. sales, or his global market share, a means of
discounting his price in a manner not easily monitored by other OPEC members.
Because of transshipment and swapping of oil, higher U.S. sales by some
governments would not be a reliable measure of aggressive bidding for tickets.
Nor could OPEC depend upon the evolution of total market shares to identify
cheaters since the monthly scramble to defend sales might obscure any long-
term trend.

The OPEC members could pledge to each other to pay no more than a certain
amount for the tickets or to pay nothing at all, but the secrecy of the
bidding process and/or the use of front companies would prohibit monitoring
compliance. Another OPEC response might be to try to buy up all the tickets
and destroy them in a high-stakes game of chicken equivalent to trying to
organize a one-month embargo against the United States. To counter this
threat if no tickets were presented for a week, the U.S. would hold the next
month's auction early with an extra three weeks maturity on the tickets.
Finally, OPEC could set up a joint sales agency to force the suspension of the

ticket sales. To be effective, the joint sales agency would have to be granted exclusive rights over all cartel oil, absolute power over the allocation of market shares, and the power to slash production until the United States dispensed with the ticket requirement.

The Adelman system does not require joint participation with other importing governments to be successful. But if it did prove successful, other consuming states (LDC and OECD) could be invited to join the U.S. import auction pool.

The import quota auction does not imply an attempt to "break" OPEC once and for all. Rather, if successful, it aims at setting in motion recurrent cycles of erosion in the cohesiveness of a group that recognizes the aggregate benefit of cooperating and can be counted upon to try periodically to regroup. The probability of its success, however, is highly sensitive to one's estimate of the amount of spare capacity held idle by governments that need the revenues such capacity could generate. Without a certain amount of unwanted spare capacity within OPEC, the motivation to bid aggressively will be low. It is not clear, however, how much unwanted spare capacity would be needed to trigger strong bidding. On the one hand, 3 mbd of unwanted spare capacity in relation to exports of 30 mbd could be absorbed if each member reduced production by only 10 percent. On the other hand, the same 3 mbd of unwanted spare capacity could mean, with the U.S. auction system, Venezuela, Indonesia, and Nigeria, for example, could have their exports to the United States completely lost, simultaneously, if they did not bid aggressively on the tickets.

The analysis presented in this study indicates that unwanted spare capacity will be large (5.0 mbd in 1980) and last a long time. The market for OPEC oil through 1985 will be the mirror image of the earlier 1973-75 period--a continuous buyers' market, with more sellers anxious to dispose of their product than there are customers at the given price. Such conditions will make a favorable environment for the successful implementation of the Adelman plan.

One should note that the Houthakker and Adelman proposals are not at all incompatible, and could be introduced in tandem to reinforce each other.

10. Whether the United States pursues an explicit bilateral agreement on oil prices with Saudi Arabia or seeks to use impersonal market forces to discipline OPEC as a whole, the challenge that the country faces in the years ahead is not only how to conserve energy at home but how to build upon that conservation effort policies that prevent OPEC from perfecting its structure as a cartel whose collective appetite for higher revenues will ultimately prove insatiable.

APPENDIX A: SAUDI ARABIA

To try to calculate how much the projects included in the Saudi Arabian
Second Plan would in fact cost now (in 1976), or in the years in which they
are carried out, this study employs the following methodology:

a) it multiplies the 1974 Plan estimates for capital projects by the
amount heavy construction costs rose in the United States between mid-1974
and mid-1977. This raises the cost-base from which the great majority of
Saudi projects was estimated to 1977 prices;

b) it multiplies the product from (a) by the difference between the
"locational premium" (between Texas Gulf Coast sites and Arabian Gulf Coast
sites) used in the Plan and a contemporary recalculation of the same "locational
premium." This tells how much the Plan projects would cost in 1977 if carried
out by U.S. companies with no further diseconomies of scale in implementation
besides those already included, implicitly, in the revised "locational
premiums." (As will be seen later, this series of cost estimates includes
expenses associated with infrastructure bottlenecks and construction delays,
but no demurrage charges. This study assumes that demurrage charges will drop
to zero by 1978.)

c) it multiplies the amount of a project that will be undertaken after
1977 by the difference between the local rate of inflation on locally procured
items and the OECD rate of inflation over the period between 1977 and the time
the local procurement is made. The local component of project costs is assumed
to be 50 percent (30 percent goods and 20 percent services). (This is close
to corresponding assumptions made by U.S. government analysts.) The local
component of recurrent costs is 90 percent.

To make the calculation required for (a)--the amount heavy construction costs rose in the United States--I have used the two authoritative published indexes mentioned in the text (Chemical Week and Nelson's) and three indexes constructed for internal use by large U.S. construction companies. On the basis of these, I take 30 percent as the base rise.

To try to determine (b)--the current "locational premium" between the Texas Gulf Coast and the Persian/Arabian Gulf Coast--I use the survey of U.S. companies and the U.S. Army Corps of Engineers mentioned in the text. The current "locational premiums" range from slightly over two times Texas Gulf Coast costs for industrial construction to slightly under three times Texas Gulf Coast costs for military construction.

In the case of (c)--the "Saudi inflation premium" above the annual rise in the OECD price index--I have used the following "local inflation premium": 31 percent for 1976/77, 20 percent for 1977/78, 10 percent for 1978/79, 5 percent for 1979/80, 0 percent for 1980/81 and thereafter.

Other Assumptions

1. Administration: this study assumes that there is 25 percent "fat" in the administrative budget, but that the continuing administration of a total expenditure program of more than $25 billion per year cannot be handled for less than 75 percent of the costs in 1975/76 ($5 billion).

The administrative budget, at whatever level, is assumed to include 80 percent local expenditures which will be subject to any Saudi "inflation premium" above the OECD price level.

2. "Other": the "other" category includes foreign aid, emergency funds, food subsidies, and general reserves. It is assumed that half of the recurrent expenditures in the "other" category are local and thus subject to the Saudi "inflation premium." The other half, principally foreign aid, is not subject to any "inflation premium" at all.

3. "Recurrent costs" are prorated, as described in the text, in accord with the amount of the Plan's projects that are being carried out at any one time.

4. "Gas and desalination": the revised figures for gas gathering and desalination projects are assumed to include an accurate estimate of a local inflation premium over the construction cycle, and to include their own "recurrent" costs.

5. Demurrage charges: this study assumes that demurrage charges will essentially be reduced to zero by 1978.

6. For project estimates under Defense, 50 percent (primarily construction and training) are assumed to be subject to the "Saudi inflation premium" and the "locational premium," and 50 percent (equipment purchases) are assumed to be subject to neither. This may significantly underestimate the post-1974 increase in price for military weapons and related support systems.

This produces the calculation, given in the text, that the original $142 billion Plan (1974 prices) would in fact cost at least $296.5 billion (1977 prices) as of mid-1977 if it could be carried out with no diseconomies of scale in implementation. If the Plan were cut in half or spread out over ten years, and the Saudi inflation rate were hypothetically reduced from its current 35-40 percent annual rate to the OECD average by 1981, Saudi expenditures would have to rise from the current level of about $22 billion per year to an annual rate of more than $59 billion (1977 prices) from the late 1970s through 1985.

Neither of these is a realistic expenditure pattern for Saudi Arabia. To represent Saudi spending choices with more plausibility, the chapter on Saudi Arabia attempted two simulations: an "immediate, drastic cutback" scenario, and a "momentum and crunch" scenario. These require annual budgets on the order of $35 to $46 billion. In each case the simulation reflects the following national priorities: domestic stability and the ability to influence intra-Arab politics (via foreign aid) = high; economic development = low; physical infrastructure and national defense = moderate.

There is no reason, of course, why Saudi Arabia has to finance its yearly budget on an annual basis. Indeed, most of the oil production projections included in this study assume that the Saudi government will draw down on its foreign exchange reserves. It can also borrow domestically or abroad.

U.S. government analysts estimate that the Saudi government had about $49 billion in official assets at the end of 1976. This includes holdings of gold and foreign currencies plus bonds and stocks held by SAMA. It is a figure lower than other estimates, which apparently include as assets foreign aid loans, e.g., to Egypt and the Sudan. This study uses the $49 billion figure as a reasonable estimate of what the Saudis will calculate they can draw upon for domestic expenditures.

A hypothetical profile of Saudi expenditures, combining the "momentum" and the "drastic cutback" scenarios might look like the following:

Saudi Expenditures
(1977 dollars)

	($ billion)
1977/78	$32.0
1978/79	37.0
1979/80	42.0
Total	$110.0

A hypothetical profile of Saudi petroleum sector revenues might look like the following:

Saudi Oil Export Revenues
(constant real oil prices)

	mbd	($ billion)
1977/78	9.5	$39.9
1978/79	7.0	29.4
1979/80	7.0	29.4
Total		$98.7

With these scenarios, Saudi reserves would peak in 1978 at $56.9 billion, and decline to a level of $37.7 billion by 1980.

Physical Import Capacity for Saudi Arabia

U.S. government studies indicate that Saudi port capacity should increase from 7,080 thousand tons per year in 1976 to 14,490 thousand tons per year in 1980. The port sample includes Dammam, Jubail, Jiddah, Jizan, and Yanbu. Air freight and road cargo capacity should increase from 895 thousand tons in 1976 to 1,334 thousand tons in 1980. This gives a total import capacity in 1980 of 15,824 thousand tons. Port births and other facilities needed to handle exports have been netted out. Total import facilities are then broken down into cement, grain, and dry cargo capacity, and multiplied by 1975 unit import values in each category to produce a constant 1975 dollar value in relation to the projected needs of both the public and the private sector. Tonnage throughput has been calculated on the basis of evidence from the full capacity years 1973-75. Any increases in efficiency by 1980 (e.g., in throughput per port birth) will result in higher import capacities than those indicated here.

Project expenditures are estimated to include 30 percent goods imports. Current and administrative expenditures are estimated to include 10 percent goods imports.

Reasons why these calculations for Saudi Arabia may be
too low:

1) The imported costs of heavy machinery and equipment are projected
here to rise no faster than the general OECD price index, whereas many
observers suggest that such items have an "inflation premium" of their own.

2) The imported costs of military equipment, including communications
and other electronic support equipment, are projected here to rise no faster
than the general OECD price index, whereas many DoD planners in program
assessment and evaluation use a figure of double the general price index
to compute expected increases for military hardware.

3) The defense figures here include no "inventory" for "front line"
Arab nations.

4) The lowering of the inflation premium to zero by 1980 is considered
excessively optimistic by many analysts.

5) The elimination of demurrage charges by 1978 is also considered
excessively optimistic.

6) For any major project, public expenditures may be bunched at the
end rather than spread out in equal segments.

Reasons why these calculations for Saudi Arabia may be
too high:

1) With increasing competition for the Saudi market, construction profits
may be reduced. It is not known if there is much "fat" to be taken off.

2) "Service fees" to Saudi procurement specialists might, hypothetically,
be reduced or eliminated.

APPENDIX B: IRAN

The methodology for the recalculation of Plan costs is given in the
text, pp. 31-32 . The assumptions made about three crucial variables are
discussed there: 1) the index of the rise in the costs of heavy construction
in Iran since 1972; 2) the rejection of the hypothesis of systematic underestimation of original program costs; 3) the escalation of the current account
budget since 1352 (1973/74). With regard to the latter, Plan and budget
office authorities set as a goal that the ratio of current to capital costs
would drop to 0.84 by 1355 (1976/77). This analysis assumes that Iran will
reach the target of 0.84 in 1977 and remain there to the end of the decade.
This assumption may be too optimistic.

Further Assumptions

4) Inflation premium in the future. For the future, this study has
adopted the Iranian PBO assumption of an average inflation of 12 percent per
year, or 6 percent above the OECD rate, to 1980. For capital expenditures,
however, one-third are assumed to represent imported equipment. Hence,
Iranian outlays for capital expenditures lose only 4 percent per year in
purchasing power due to the local "inflation premium" above the OECD rate.

5) Petroleum production capacity. It is assumed that the Khuzestan
Fields will reach a "theoretical" capacity of 6.5 mbd, and an actual operating
capacity of 6.2 mbd by early 1977 and maintain this capacity for at least
eight to nine years. Operating capacity elsewhere will reach 0.8-0.9 mbd
by 1980.

6) Domestic petroleum consumption. The following figures have been used
for domestic petroleum consumption: 1976/77 = 0.4 mbd, 1978/79 = 0.5 mbd,
1979-80 = 0.5 mbd.

7) <u>Petroleum export capacity</u>. 1976/77 = 6.4 mbd, 1977/78 = 6.4 mbd, 1978/79 = 6.5 mbd, 1979/80 = 6.6 mbd.

8) <u>Natural gas exports and revenues</u>. 1976/77 = 321 million cubic feet, $0.2 billion; 1977/78 = 321 million, $0.2 billion.

It appears that Iranian planners tend to underestimate their petroleum reserves, and to overestimate the export potential of natural gas, steel, and copper in the future. The figures that are used for Iran's oil reserves come from company calculations that have been generated in the past in an era of low production costs and low crude prices. While they may be accurate for oil at a cost of 12¢ per barrel, they are hardly accurate for the amount of oil that could be recovered for costs ranging from $1 to $10 per barrel. If the climate were made favorable for foreign companies to mount extensive new exploration efforts in Iran, current market conditions would almost certainly result in large additions to the estimate of the country's reserves. This would require, however, a rather dramatic reversal of Iran's approach, which has reduced the margin on foreign company liftings from $0.40-$0.80 in the 1960s to $0.22 at the present time. (On this point, see Fereidun Fesharaki, "Iran's Oil Reserves: A Reevaluation," The Institute for International Political and Economic Studies, Tehran, processed.)

With regard to natural gas exports, there is no doubt that the country has large reserves. It is a subject of much controversy in Tehran, however, whether these can be developed fast enough to meet the commitment to the Soviet Union for IGAT II by January 1, 1981. IGAT II would approximately double Iran's export level. The country currently earns approximately $0.3 billion per year from gas exports. Over the longer run it probably makes more economic sense to use the country's natural gas for local power (and fertilizers, etc.) rather than exporting it, with the goal of reserving petroleum for export. Iran presently earns about $2.00 per million Btu for crude exports and $0.20-$0.30 per million Btu for gas exports.

Most careful analyses of Iran's steel export possibilities suggest that all of the country's production for the foreseeable future will be needed internally. With regard to copper, most of Sar Chesmeh's initial production in the late 1970s and early 1980s (145,000 metric tons) will be exported with new additions to capacity more than making up for the increases in domestic demand. Over the longer term other copper projects are quite possible.

The net result of these calculations suggests that the long term decline of Iranian exports after 1984, while still inevitable, could be made more gradual than some current government projections indicate.[5]

9) Non-energy tax revenues. This study adopts the PBO assumption of a growth of tax revenues from sources other than petroleum and natural gas of 13 percent per year from a base of $3.9 billion in 1975/76.

10) Government revenue per barrel. Government revenues are calculated on the basis of a 22¢ per barrel discount to the consortium and a 30¢ per barrel production cost (including amortization of development expenses). While this may accurately represent the country's "take," it may overstate the revenues accruing to the central government since NIOC administrative costs and investment expenditures must come of this "take."

11) Defense expenditures. There is no breakdown of Defense expenditures into "current" and "fixed investment" categories in either the Plan or the

[5]Moshen Fardi, "Iran's International Economic Outlook," in Jane Jacqz, ed., Iran: Past, Present, and Future (New York, Aspen Institute for Humanistic Studies, 1976).

budget. In addition, as explained in the text, U.S. government analysts argue that there is a large amount of military construction included in categories other than "Defense."

12) Foreign aid. Iranian projections for foreign aid dropped 41 percent between 1975/76 and 1976/77, from $2.5 billion to $1.5 billion. It is assumed here that the $1.5 billion level will hold steady to 1978.

13) Annual deficit and foreign borrowing. An annual deficit of $3.0 billion has been included in the first budgets simulated for 1976/77 and 1977/78 (based on exports of 4.6 mbd), financed by a combination of foreign and domestic borrowing.

14) Demurrage and port delay charges. Demurrage charges amounted to about $500 million, and other port delay and congestion charges amounted to another $400-$500 million in 1975/76. It is assumed (too optimistically in the view of many observers) that such charges will be reduced to zero by 1978.

Projections and Simulations: Iran ($ million)

(1) Public fixed investment in the original Fifth Plan ($1 = 67.50 rials) in million dollars:

> Economic Affairs $15,809
>
> Social Affairs   5,333
>
> Public Affairs   1,801

(2) Public fixed investment in the original Fifth Plan with implicit 5 percent per year inflation assumption squeezed out; that is, 1351 (1972/73) price of the original capital programs.

methodology: (1) has been divided into five equal real spending segments from which a 5 percent per year inflation component has been factored out (i.e., for the third year the inflation component

is 15 percent).   The five real spending segments are then
summed.

Economic Affairs $13,747

Social Affairs   4,637

Public Affairs   1,566

(3) Capital expenditures in current figures for 1352 (1973/74) and 1353
(1974/75). ($ = 67.50 rials)

1352

Economic Affairs $1,367

Social Affairs   527

Public Affairs   495

1353

Economic Affairs $3,172

Social Affairs   1,234

Public Affairs   760

(4) Progress in 1351 prices represented by 1352 and 1353 capital expenditures.

methodology: (3) deflated by 21 percent for 1352, by 51.3 percent for 1353.

Economic Affairs $2,625

Social Affairs   1,018

Public Affairs   761

(5) Capital projects left over from original Fifth Plan as of the beginning
of 1354 (1975/76) in 1351 prices.

methodology: (2) minus (4).

Economic Affairs $11,122

Social Affairs   3,619

Public Affairs   805

(6) Cost of remaining capital projects from original Fifth Plan as of the
beginning of 1354 (1975/76) in 1354 prices.

methodology: (5) x construction cost escalation since 1351 (181 percent).

        Economic Affairs $20,198

        Social Affairs   6,572

        Public Affairs   1,462

(7) New capital projects added in revised Fifth Plan.

methodology: revised total public fixed capital projects minus original
total fixed capital projects ($1 = 67.50 rials) (million
dollars).

        Economic Affairs $28,319 - 15,809 = $12,510

        Social Affairs   8,238 - 5,333 =   2,905

        Public Affairs   5,638 - 1,801 =   3,837

(8) New capital projects added in revised Fifth Plan with implicit 12 percent
per year inflation component squeezed out. This gives the costs of the new
capital projects in 1353 (1974/75) prices.

methodology: (7) has been divided into three equal real spending segments
from which a 12 percent per year inflation component has been
factored out (i.e., for the second segment the inflation
component is 24 percent).

        Economic Affairs $10,171

        Social Affairs   2,362

        Public Affairs   3,120

(9) Cost of new capital projects in 1354 (1975/76) prices.

methodology: (8) x escalation in construction costs 1353 to 1354 (125 percent).

        Economic Affairs $12,714

        Social Affairs   2,953

        Public Affairs   3,900

(10) Cost of uncompleted capital projects in the revised Plan as of the beginning of the revision period (1354) in 1354 (1975/76) prices.

methodology: (6) plus (9).

Economic Affairs $32,912

Social Affairs 9,525

Public Affairs 5,362

(11) Cost of remaining capital projects as of beginning 1355 (March 1976) in 1355 prices.

methodology: (8) x escalation in construction costs 1354 to 1355 (118 percent hypothesized) minus 1354 expenditures.

Economic Affairs $38,836 - 4,432 = $34,404

Social Affairs 11,240 - 1,775 = 9,465

Public Affairs 6,327 - 1,255 = 5,072

(12) Cost of all capital projects in Fifth Plan in 1355 (1976/77) prices.

methodology: (2) x escalation in construction costs since 1351 (1972/73) plus (8) x escalation inconstruction costs since 1353 (1974/75).

Economic Affairs $24,460 + 15,002 = 44,462

Social Affairs 9,937 + 3,484 = 13,421

Public Affairs 3,356 + 4,602 = 7,958

(13)	Percentage of capital projects left to complete as of March 1976	% of capital projects completed as of March 1976
	(11)/(12)	100% - (11)/(12)
Economic Affairs	77.4%	22.6%
Social Affairs	70.5%	29.5%
Public Affairs	63.7%	36.3%

(14) Total cost of rest of Plan (economic, social, and public affairs) if ratio of current to capital costs is 0.84 (in 1355 or 1976/77 prices). Note: this does not include defense or foreign aid expenditures.

methodology: (11) x 1.84.

          Economic Affairs $63,303

          Social Affairs   17,416

          Public Affairs   9,332

             Total $90,051 million

(15) Total cost of entire Plan (economic, social, public affairs) (in 1355

or 1976/77 prices for uncompleted portions). Note: this does not include

defense or foreign aid expenditures.

methodology: (14) plus total spent so far on economic, social, public

          affairs. Total spent so far ($1.00 = 67.50 rials for 1352

          and 1353; $1.00 = 69.25 rials for 1354).

          1352 (1973/74) $5,127

          1353 (1974/75)   14,846

          1354 (1975/76)   15,906

                      $35,879

          Grand Total $125,930 million

First Simulation

    Oil production averaged 5.4 mbd in 1355 (1976/77) and is hypothesized

by the Iranian Central Bank at 4.6 mbd for 1356 (1977/78); fiscal deficit

equals $3.0 billion each year; foreign aid equals $1.5 billion each year;

Iranian inflation rate equals 12 percent per year (vs. OECD rate of 6 percent);

oil export prices for 1977/78 are $12.81 per barrel for light, $12.49 per

barrel for heavy through December 1977; exports are 50 percent light/50 percent

heavy.

(16) Iranian revenues ($ billion) (1355 dollars).

1355 (1976/77)	oil	$22.3
	gas	0.2
	other	4.4
	Total	$26.9
1356 (1977/78)	oil	$20.7
	gas	0.2
	other	4.9
	Total	$25.8

(17) Iranian expenditures: these revenues would essentially support an Iranian budget of the dimensions of the 1355 (1976/77) budget ($29.7 billion) with a fiscal deficit of $3-4 billion each year.

The capital expenditures in the 1355 budget were projected at:

Economic Affairs $6,319

Social Affairs   1,850

Public Affairs   1,939

(18) Assuming a 12 percent per year Iranian inflation rate, a 6 percent per year OECD inflation rate, with one-third capital expenditures assumed to be imported equipment, the Iranian capital budget loses 4 percent purchasing power in 1356 in comparison to 1355 prices. Capital spending (1355 + 1356) in 1355 prices.

Economic Affairs $12,385

Social Affairs   3,626

Public Affairs   3,800

(19) Percentage completed by end of Plan period if Iranian oil production averages 4.6 mbd to end of Plan period.

100% - $\underline{/}$(11) - (18$\underline{)}$$\overline{/}$ / (12)

            Economic Affairs 50.5%

            Social Affairs   56.5%

            Public Affairs   84.0%

Second Simulation

Oil production climbs to full capacity to end of Plan period with 15 percent price rise in 1977; all other assumptions the same. Export capacity is 6.8 mbd.

(20) Iranian revenues ($ billion) (1355 dollars).

            1355 (1976/77)   oil $22.3

                               gas   0.2

                               other   4.4
                               Total $26.9

            1356 (1977/78)   oil $31.2

                               gas   0.2

                               other   4.9
                               Total $36.3

(21) If Iran did not run a fiscal deficit, it could complete the following percentages of its programs (using same methodology as in first simulation).

            Economic Affairs 53.1%

            Social Affairs   59.1%

            Public Affairs   88.6%

Reasons why these estimates of the "financial squeeze" for Iran may be too low:

1) This analysis assumes that all estimates of program costs were accurate at the time they were made. Instead they may have been systematically underestimated by 30-40 percent, or more, especially in the more remote parts of Iran (including Bandar Abbas and Chah Bahar).

2) The escalation of capital costs may have been substantially higher than the estimates used here.

3) The revenue per barrel available to the Iranian government to finance public programs may be lower than the 1977 average "take" of $12.13-$12.74 per barrel hypothesized here due to NIOC administrative and investment charges that must be subtracted.

4) The Iranian "inflation premium" for major government programs and projects in the future may be more than the 6 percent per year assumed here.

5) The PBO assumption (adopted here) of an annual increase of 13 percent in non-energy tax revenues may be too high.

Reasons why these estimates of the "financial squeeze"
for Iran may be too high:

1) The Iranian government may be able to push the ratio of current to capital expenditures below 0.84.

2) The repressurization of Iranian oil fields may prove more successful by the late 1970s and early 1980s than currently supposed. This could ease Iran's revenue shortage, but intensify the balancing problem for OPEC.